Dr Lynne Kelly AM is an educator and an Adjunct Research Fellow in the School of Humanities and Social Sciences at La Trobe University, and an Honorary Fellow in the School of Physics at The University of Melbourne. Over the past decade, she has focused on Indigenous knowledge systems and the use of story, song, place and art as mnemonic devices. She is an Australian Senior Memory Champion, and her memory work is used widely in schools and universities. She is the author of twenty books on science, including the bestselling *Memory Code*, *Memory Craft* and *Songlines*.

www.lynnekelly.com.au

———

'At last, here is the Big Picture about what it really means to be human!

The present is catching up with our past. The discovery of how the supergene works is the key to how First Peoples can remember knowledge across 2000 generations without the written word. Humans are wired to learn and remember through story telling conveyed by immersion in art, music and performance. This embodied knowledge system is integrated, in contrast to the Western system of text-based compartmentalised disciplines that bypasses the supergene.

This is a call to action on how to learn the way we are supposed to.'

Adjunct Professor Margo Ngawa Neale, Centre for Indigenous Knowledges, National Museum of Australia

T0362925

'A consensus is growing in the world, amongst a scattered family of scholars and knowledge keepers from many cultures: that consciousness, place, story, craft and memory are bound together to form an extended self with a unique neurology embodied far beyond the limits of our skulls. Lynne Kelly is a leading light in a global community of intellectuals, whose regalia may be academic robes or grass skirts, who think with their hands, feet, homes and hearts, making marked paths of memory for others to follow. Kelly rounds out an expansive trilogy that began with *The Memory Code*, offering a map anyone can follow to claim an ancestral legacy of genius held in trust within the body of every human being.'

Tyson Yunkaporta, Indigenous Knowledge Systems Lab, Deakin University, author of *Sand Talk*

'A rollicking ride through the strange and wonderful world of how humans attained the ability to learn, store and transmit vast amounts of information, invigorating communities with memory, knowledge, meaning and culture through the generations.'

Professor Bruno David, Monash Indigenous Studies Centre

'Dr Lynne Kelly is a veteran of authoritative books on memory. Now she adds music and art and their connection to memory and genetics. Along the way she surveys the archaeological evidence. All of this is written in an easy-reading style to sweep us along in the tide of an engrossing story. Written in close connection with the geneticists and others who ensure that the book is authoritative this is a book full of fascinating details of so many aspects of genetics, evolution, music, art and knowledge all driven by Kelly's deep understanding of memory and oral tradition.'

Iain Davidson, Emeritus Professor, Archaeology, University of New England

'With its uncompromising title, Lynne Kelly offers us a thought-provoking book on human evolution, emphasising the critical role of creativity and art throughout our history, and their continued importance in promoting individual and societal wellbeing in the 21st century.'

Professor Alan Harvey, author of
Music, Evolution, and the Harmony of Souls

'Whether or not you subscribe to a "knowledge gene", let's all rejoice in the unique human genome that gives us language and knowledge, and allows us to store, retrieve and transmit knowledge over time and space. I really enjoyed reading a book so upbeat about the genes that make us human.'

Professor Jennifer A. Marshall Graves FAA, AC,
School of Life Sciences, La Trobe University

'Lynne Kelly takes us on a mnemonic *tour de force*, exploring groundbreaking neuroscience research that reveals how our ability to develop knowledge and commit it to memory through story, music, and art is encoded in our very DNA. *The Knowledge Gene* changes everything.'

Associate Professor Duane Hamacher,
author of *The First Astronomers*

'Lynne Kelly walks us from the ancient into the modern world and right into our classrooms, providing the research basis and lived experience of how music and the arts enrich—and inevitably create—our human condition.'

Dr Anita Collins, author of *The Music Advantage*

'*The Knowledge Gene* is an incredibly important piece of work which brings to light information that will forever change the way we think about our history, how we learn, and what it means to be human.'

Nelson Dellis, five-time USA Memory Champion

Other books by Lynne Kelly

Crocodile: Evolution's greatest survivor
Exploring Chaos and Fractals (co-authored)
Knowledge and Power in Prehistoric Societies
The Memory Code
Memory Craft
The Skeptic's Guide to the Paranormal
Songlines: The power and promise (co-authored)
Songlines: First Knowledges for younger readers
(co-authored)
Spiders: Learning to love them

The Knowledge Gene

The incredible story
of the supergene
that gives us
human creativity

LYNNE KELLY

ALLEN&UNWIN
SYDNEY • MELBOURNE • AUCKLAND • LONDON

First published in 2024

Copyright © Lynne Kelly 2024

Allen & Unwin
Cammeraygal Country
83 Alexander Street
Crows Nest NSW 2065
Australia

Phone: (61 2) 8425 0100
Email: info@allenandunwin.com
Web: www.allenandunwin.com

Allen & Unwin acknowledges the Traditional Owners of the Country
on which we live and work. We pay our respects to all Aboriginal and
Torres Strait Islander Elders, past and present.

 A catalogue record for this
book is available from the
National Library of Australia

ISBN 978 76147 070 7

Set in 13/17.6 pt Bembo Std by Midland Typesetters, Australia
Printed and bound in Australia by the Opus Group

10 9 8 7 6 5 4 3 2 1

 The paper in this book is FSC® certified.
FSC® promotes environmentally responsible,
socially beneficial and economically viable
management of the world's forests.

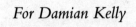

For Damian Kelly

CONTENTS

FOREWORD
BY DAVID KANOSH

'*From time immemorial, we who are Tlingit have lived upon these shores. These are the stories of old for these are the stories which go back to the time of legend.*'

I can still hear my grandparents saying those very words before they told me some stories. These stories were often associated with places throughout south-east Alaska. As we rode the ferry from Angoon to Sitka, they'd bring my attention to a location on a nearby island, highlighting features on the landscape, and I knew a story was about to begin.

One of my favourite stories is of Raven fighting his uncle, who had tried killing Raven by various means, and finally causing the tides to rise and summon a storm to kill Raven. The result was a flood, which sent the Tlingit people running up mountains and jumping into canoes to escape or ride out the flood.

My grandparents had taught me these mnemonic techniques to elaborately encode into long-term memory vast amounts of information in a relatively short time. They had taught me other techniques as well, but those techniques are reserved for only a select few. People who are familiar with the memory palace technique and the use of *loci* will find some similarities with what I had been taught.

The beautifully carved totem poles, wooden carved ceremonial hats and even the wooden carved rattles. They are more than art. They serve a very practical purpose. Each had places or *loci* that I had used to elaborately encode information.

My grandmother picked up a piece of plywood that she had glued pebbles to and said simply, 'You will remember.' She then touched each piece as she recounted stories. To my young mind, what she was doing seemed like magic!

I made my own memory boards. As I glued each piece to the plywood, I thought about the stories I would tell with this memory board. I had spent an afternoon just putting on five pebbles! How did my grandmother make it look so easy? Practice. She'd been at this for a while.

My grandparents had smiled. They were happy I was taking such an interest in wanting to learn the stories which had been handed down from our ancestors to us.

I had been taken to some rock cairns which are atop some of the mountains in south-eastern Alaska. What was this? Rocks in those rock formations were used to help tell part of a story, and each rock cairn had different stories associated with it.

After being taken to the rock cairns, I had been fascinated by other rock formations from around the world, such as Stonehenge. I had always wondered: who were the storytellers that used Stonehenge? What knowledge was shared? The decades passed but my curiosity about rock formations found around the world never went away.

On numerous boat trips, I have gone with friends to these places where the stories unfolded. I see where Raven and his uncle were fighting it out and causing a flood. The very landscape serves as a memory aid for me to remember this story, but that's just part of it. I don't just tell a story, I act it out. I dance it. When I'm alone, I sing it.

I stand upon the land where my ancestors have stood. I stand on the land where untold numbers of storytellers have stood. I take in the view of the landscape. I close my eyes for a moment as the story becomes alive again within me. It's now time to bring that story out.

I open my eyes. I look at my audience. I echo the words which were spoken to me by my grandparents so long ago: 'From time immemorial, we who are Tlingit have lived upon the shores . . .'

The story begins. The land is my stage. Who am I? I am Raven fighting his uncle. I am Raven's uncle fighting Raven. I am the Tlingit people running up the mountain to escape the flood. I am the Tlingit people jumping into canoes to ride out the flood.

Dr Lynne Kelly's book *The Memory Code* had been brought to my attention by Josh Houston at a potluck before a meeting. As Josh explained what the book was about, my curiosity was becoming stronger and stronger. The moment Josh said 'Stonehenge', I had to find out more. That was it. I had to buy the book.

As I read this book, I stood up not a few times, saying, 'Yes! Yes!' I cried a few times. I've never really read anything written about our people. Not in English, anyway. She was writing about orality, the manner by which orally based traditions are used to memorise vast amounts of information in a relatively short amount of time. I had paused quite a few times, reflecting on everything which my grandparents had taught me.

I had shared information with anthropologists before about what I know but none of them had ever asked how I had so elaborately encoded what I do know into long-term memory. I had been impressed with what Dr Kelly had written about. I added her other books to my collection. *The Memory Code*, *Memory Craft* and *Songlines: The power and promise*. Was that it? Had I reached the end of the journey? Far from it.

I have shared about the culture of my people, the Tlingit, and some of the various mnemonic devices with Dr Kelly. This has been a learning journey for both of us. It's a wonderful journey. As much as I have been a teacher, I've also been a student. There is so much to learn. That is the magic of it! Our minds do not have a 'use-by date'.

Dr Kelly has invited you and me to be a part of the journey ahead. The journey continues. The story continues.

There's still a bit of magic in me. I am Raven flying up to the sky. I am every Tlingit storyteller who has ever lived. I am the student who is eager to learn. I call the end back to the beginning!

David Kanosh
Yookis'kookéik
Tlingit Elder and storyteller,
South-Eastern Alaska

LIST OF FIGURES

CHAPTER 1

Knowledge
makes us human

You belong to the only species that can learn and share what you have learnt in ways that create an ever-growing encyclopaedia. We can constantly learn and adapt by drawing on new information without losing what we have already learnt. That has given us an advantage over every other species—an advantage we have not always used well.

Our closest relation genetically, the chimpanzee, does not produce art, does not perform music, does not tell stories and debate philosophy, and has not adapted to environments beyond its natural habitat. The differences between us and chimpanzees are in our genes, within the 1.2 per cent of coding DNA that separates us from them. This book is all about a mutation that separated us from the chimpanzees hundreds of thousands of years ago and supercharged our ability to store information in our uniquely human way.

You have two copies of this ancient mutation, but it isn't an entirely rosy story. This mutation also introduced the potential for a devastating disorder, one that has never been recorded in chimpanzees. Given that risk, what was so valuable that evolution left this mutation in our gene pool in the first place?

Only very recent discoveries can answer that question. And it's the story of those discoveries that this book will tell.

In September 2021, my life was blissfully full: I was starting new projects, consolidating old ones and tying up innumerable loose threads. The last thing that I needed was that email, that new word.

The time had come to indulge in all the treasured plans impatiently waiting on the backburner. During the previous decade, I had become fascinated by the way Indigenous knowledge keepers managed to remember so much information about their environment when, right through school, I'd struggled with anything that relied on memory. My logic was strong, so mathematics was a dream. My memory was pathetically weak, though, so foreign languages were impossible.

How could Indigenous Elders possibly achieve their incredible feats of memory? Despite the distinct nature of the many Indigenous cultures around the world, they all use remarkably similar methods to recall vast amounts of information. They have memorised the names and uses of thousands of plants and the behaviour of every animal in their traditional lands, including even the multitude of insects. They could navigate hundreds of kilometres and name all the stars. They knew their relationship to every person in their community and beyond, and to every place within their landscape. They maintained laws and ethical expectations, songs, dances and stories. And it was all kept in their memory. How the hell did they do it?

Over a decade after first asking that question, I had written a doctoral dissertation and five books on the topic. I was enthralled by the methods they were using. I had started experimenting with the techniques and was shocked at how effective they proved to be. My natural memory was still poor, but I could now train myself to memorise almost anything.

So I did. Every country in the world, a timeline for all of history, a field guide to all 412 bird species in my state, a thousand digits of pi . . . I kept getting excited about new experiments and different methods. I took on French, having failed it at school. And when I found that even that was within my grasp, I decided to take on my ultimate challenge and learn Chinese. I enrolled in French classes, live and online. I enrolled in Chinese classes, live and online, and started sessions with a friend who had spoken it since she was a child.

The memory methods used space, large spaces across the landscape and small spaces in handheld memory devices. I had over ten kilometres around my neighbourhood mapped out and encoded with the minimum I needed to test the methods. I had already memorised lots of information, but I wanted to fill my memory spaces with all the data I'd carefully copied and pasted in my numerous notebooks. I wanted to layer these memory spaces with more and more complex information, the way Indigenous Elders do.

The methods use art in wondrously varied forms. I had created a few mnemonic artworks to test the effectiveness of this, but there were so many more scribbled plans in yet more notebooks. I had become obsessed with Chinese handscrolls, a painted narrative form that seemed to me to be cognitively the same as Australian Aboriginal sand talk. I planned scrolls telling stories of birds and of Shakespeare. I bought art materials. I set up my studio, moving the tools of my writing career aside to make room for paints and paper. I enrolled in art classes, drawing and painting live, and calligraphy online. I spent the most exhilarating time with my new mentor, a 101-year-old artist who is still painting and still exhibiting. I knew I would learn a lot more from her than just about watercolour.

And I wanted to know more about music, especially as rhythm is so critical to non-literate knowledge systems. Indigenous cultures all engage with music. The Northern European Sámi drums are covered in markings that look very like the mnemonic marks I had studied elsewhere. Being close to tone-deaf, I had put my music experiments on that rather overloaded backburner and hadn't even started making notes. Western music sounds so different from Chinese, and then there's Indian and African. Yet every culture I had explored used music as a fundamental tool in their way of storing knowledge. Why was the use of music universal and yet the sound of it was not? My lack of engagement with music was holding me back, and this was something I was going to fix. I had taught the physics of music often but had little experience of the real thing.

So I started on a music course with the help of an opera-singing, physics-loving friend. Opera is a narrative form, so it seemed relevant. But I really didn't like opera. Still, I was willing to endure it for the sake of my research. I watched *La Traviata*. It wasn't as bad as I had expected. And the second time through, it was actually quite good. By the third time, I was besotted.

I tried *The Magic Flute*. Early on there is a trio of women screeching, and then the Queen of the Night screeches even more. I skipped the screeching because I was starting to love the rest. But to engage properly, I told myself, I had to endure, so I stopped the skipping. By the third time listening to the screeching, those arias had become my favourite pieces of music. I didn't know what was happening to my brain, but in my year of bliss, I was going to try to work it out as I explored different sounds and rhythms from across the world.

In my previous books I had shown how ancient monuments the world over, from the European stone circles to the amazing mounds across the Americas, the Pacific islands statues and the Australian painted rock shelters, were all designed in ways that would enhance the storing of knowledge. There were so many more monuments across Europe that seemed to conform to the same pattern, but I had yet to analyse them properly. I was most excited about the Scandinavian Neolithic stone ships; I just needed time to go through the archaeological reports. And there were drawings of stone circles in India in a really old book I'd found but never had the time to read. More notebooks listed hundreds of other sites that were demanding my attention. And I desperately wanted to give it.

I had stopped training for memory competitions. A few years before, I had taken out the Australian senior memory championships. I admit, there weren't many people competing in the over-60 category, but I had won. Twice. I wanted to train again.

There were only 24 hours in the day, but every one of them, less a few for sleep, was to be filled to overflowing with things I longed to do.

And yet all the time I had a sense that there was something more, that I was only glimpsing the potential of the memory technologies that I had identified. I felt that there was a much broader and more ingrained potential, but I had no idea how to find it.

A message, among many that day, was the catalyst, sending me on a journey that would far outstrip anything I could have imagined.

A stranger adds *neurofibromatosis* to my vocabulary

Dr Andrea Alveshere contacted me from Illinois in the United States, wanting to send me some information about her research. I agreed and a long email arrived the next day. I noticed her epistle in my inbox, but it was a beautiful morning and I wanted to walk the long way to art class. So I ignored it. That afternoon I glanced at it. It seemed a bit complicated, so I decided that I could ignore it a little longer and headed to the studio.

Yet I couldn't get her words—those I had registered—out of my head. It seemed to be about a genetic discovery. And something about evolution. But my genetics knowledge was sketchy and years out of date, and my evolutionary knowledge lacked sufficient depth.

I did not need any new ideas. I had notebooks full of them. Even from that skimmed read, I knew that to engage would force me to re-examine all the work I had done and all those entangled threads and try to make sense of the whole thing through two completely different lenses. That would shove all my wondrous dreams back onto that crowded backburner.

I did not want that email.

I had written that the only thing that all the different cultures had in common was the human brain. Indigenous cultures, separated by vast spans of space and time, did not teach their incredibly similar memory tools to one another. They must all use the same methods for innately human reasons. There was neuroscience to explain why associating memory with space was so effective. But why did all cultures use art and music as fundamental aspects of their knowledge systems? Why did all cultures implement performance-based knowledge systems when they didn't use writing? I had put those questions on the backburner as well. Right at the back.

Did some obscure genetic discovery explain the connectivity that I felt had to be there? Something that linked all of these apparently disparate aspects of human knowing? Did they really date back to an event hundreds of thousands of years ago? Could I ignore an email that offered a possible solution? And was I willing to sacrifice my year of bliss?

I couldn't ignore her. A biological anthropologist was writing to me about the evolution of a gene that I had never heard of. When the gene was corrupted, it led to an equally unfamiliar disorder, neurofibromatosis type 1. I had to practise saying that quite a few times before it rolled off the tongue. Fortunately, it is usually referred to as NF1 (more correctly written as *NF1*).[1] But what did this have to do with knowledge?

Dr Andrea Alveshere is an Associate Professor in Anthropology and Chemistry at the Western Illinois University. I started to glimpse the reason she was writing to me when she mentioned ancient art. She talked about how intriguing she found the parallels between the appearance of the earliest art and the spread of culturally modern humans across the globe. My work had led her, she wrote, to understand the mechanism by which such knowledge was stored. Fair enough. I was with her so far.

Andrea and her research collaborator—one of the world's leading experts on NF1, Dr Vincent Riccardi—had been exploring the impact of this disorder on many skills, which, intriguingly, included art, music, storytelling and connection to place. These were key words in all my work on memory systems. Andrea's long email went on to explain that she had identified a uniquely human variant of this gene, demonstrating that it was part of the tiny suite of DNA differences between us and chimpanzees. I had never thought about chimpanzees. Clearly, they don't make art and music. I didn't know about their spatial skills.

I had traced signs of the memory techniques back to the British Neolithic, a mere 5000 years ago. In Australia, the evidence dates back for tens of thousands of years. Andrea seemed to be saying that the knowledge systems I had identified were closely related to a gene mutation way back when our human lineage separated from that of the chimpanzees. Our last common ancestor was at least 600,000 years ago. That's way before the Neolithic.

After reading Andrea's email a few times, I had two thoughts: 'Wow, this is amazing!' and 'This is way too speculative for me'.

I sent the email to a dozen of my academic colleagues in various fields. The responses came back universally encouraging me to engage with Andrea. They did warn me against any single-gene theory and pointed out some major leaps in logic, probably due to her need to keep the explanation as brief as possible. Over the following months, as Andrea addressed these concerns, I became convinced that the hypothesis was well worth exploring further with her.

And it was even worth stacking all my other plans on the backburner again.

What is the NF1 disorder and why is it important?

The NF1 disorder is surprisingly common, affecting approximately one in every 3000 babies born anywhere in the world. That is more than cystic fibrosis and muscular dystrophy combined.[2] Yet very few people seem to have ever heard of it. In a few quick searches online, I found a number of articles about genetic disorders that I had read in the past which mentioned neurofibromatosis, and yet I had never noticed it. Could that be because it is such a difficult word to recognise, let alone pronounce?

There are many children born with NF1 in almost every sizeable community. They grow to be adults. For some the impact of the disorder is devastating, while others lead almost normal lives.

We all have two copies of the NF1 gene, one from each biological parent. Sometimes one of those copies will be corrupted, due to some kind of error when DNA was being replicated during cell division. Any mutation in either of the two copies will render that gene non-functional. The effects of the non-functional gene will dominate, so anyone born with only one fully functioning copy of the NF1 gene will have the potentially tragic disorder. The symptoms that emerge can be devastating, both medically and socially. The impact can be life-altering and even fatal.

I have come to believe that this disorder is critically important for understanding how we can all gain so much more from our genetically granted potential for learning. It is all to do with how we know what a gene provides. Genetically driven disorders occur when a gene no longer does the job it has evolved to do. By examining what has gone wrong, we get a better insight into what the gene does when it is functioning correctly.

What can go wrong when one of a person's NF1 genes is mutated points to a whole swag of traits that, as Andrea could see, were directly related to the work I was doing on memory systems in oral cultures. The more I understood, the more significant the implications became for education and learning throughout life for everyone alive today.

But it took me a while to comprehend the link, mainly because the most significant medical issues for those with NF1 are all to do with tumours. Once I had a better understanding of that devastating impact, I could concentrate on the cognitive effects, which allowed me to see the connection Andrea had made.

Mutations are not all the same, and neither is the disorder

Everyone who has the NF1 disorder will exhibit some of the symptoms described below during their lifetime. This is an unpredictable disease with a great deal of variation in the type and severity of symptoms. It isn't even predictable within families. Siblings with exactly the same NF1 genetic mutation can have vastly different outcomes over their lifetimes.

London-based identical twins Adam and Neil Pearson both have NF1. Actor and broadcaster Adam is genetically identical to Neil, but their symptoms are vastly different. Adam's face is covered by benign tumours—called neurofibromas—that are growing out of control and could cost him his sight. For this, he was bullied and called names at school. Neil has no visible tumours but has suffered from a mysterious memory loss since his teens, so severe that he cannot remember what day it is. He also has epilepsy. Their heartbreaking story has been told in a number of documentaries. You can see the impact in a screen-shot from their BBC Two Horizon documentary *My Amazing Twin* as shown in Plate 1.1.

The NF1 gene is a tumour suppressor gene, hence the asso-ciated disorder is often referred to as a tumour disorder, because the most obvious, dangerous and distressing symptoms come from tumours.

The NF1 gene serves to encode for the protein neuro-fibromin. One of the key functions of the protein is to keep our neurons—the nerve cells—growing and dividing at exactly the right rate. With one of the genes no longer function-ing properly, the production of neurofibromin is impacted. Without enough of the protein to control them, the neurons can grow out of control, producing the tumours that are characteristic of the disorder.

Sometimes the tumours appear as they do on Adam Pearson, as shown in Plate 1.1. Sometimes they are many individual growths, as shown in Plate 1.2. These can grow anywhere in and on the body, including on the face.

Neurofibromas rarely appear before puberty. Although usually benign, about 15 per cent turn malignant, possibly leading to cancer. Neurofibromas sprout from the nerve sheaths, a layer of tissue that surrounds and protects the nerves. There is currently no way to predict where the tumours will occur on the vast nervous system, nor when they will occur or how quickly they will grow.[3]

Unfortunately, surgical removal involves a high risk of nerve damage and paralysis. Removal often causes the tumours to grow back even more vigorously. Consequently, neurofibromas are only removed if they become painful or impact on vital organs. The tumours can develop in organs, causing serious health problems by pressing on nearby body tissue, including in the brain or on spinal tissue. Neurofibromas on auditory nerves can lead to hearing loss.

There is also the psychological impact of neurofibromas. Social interactions can become nightmarish, as most of the population do not recognise the tumours for what they are. Sufferers are often shunned, either because of their appearance or from an incorrect fear that they have something contagious. Sadly, those with highly visible neurofibromas often hide at home.

What causes the NF1 disorder and its potentially devastating impact?

Neurofibromin 1 is the formal name for the NF1 gene, which is found on chromosome 17. It is about 300,000 base pairs long. You may be familiar with those bases being represented by the four letters G, C, A and T when you see models or drawings of DNA in its double helix.

As I mentioned above, you have two copies of NF1, one from each of your parents. If one of those copies had been broken through mutation, then you would automatically have the NF1 disorder. If both your copies had been broken at conception, then you would not have survived long enough to make your presence known as a foetus.

About 50 per cent of mutations occur spontaneously in the sperm or egg from a parent. Consequently, this is a completely unpredictable disease and can arise in any family, even when the disorder has never occurred in that family before. The other 50 per cent are inherited from a parent with the disorder. If there's a disruption in any letter in the expected sequence anywhere along that 300,000-letter code, the gene won't make the right protein. Even with different corrupted sequences, the result is the neurofibroma tumours and the same kinds of cognitive challenges.

Everyone born with a mutation in the gene, whether inherited or spontaneous, has a 50 per cent chance of passing the NF1 syndrome on to each of their children. If they have a child who inherits the fully functioning gene from them and a fully functioning gene from their other parent, the child will not have NF1. Also, that child can never pass on the disorder. Inheriting NF1 cannot skip a generation.

Although not great odds, it is at least reassuring that about 60 per cent of those people with NF1 will end up with a minor case. Many will have families of their own.

There is currently no cure for NF1.

Some physical symptoms can appear quite early

As tumours don't usually appear before puberty, babies born with NF1 rarely have them. Often one of the first symptoms is the appearance of flat, brown birthmark-like spots on the skin of a young child. These are usually light brown, like the colour of coffee with cream, hence the name: café-au-lait spots. For a diagnosis of NF1, a paediatrician will look for six or more café-au-lait spots wider than five millimetres on a child's body. Another early sign is freckles in unusual areas, such as the groin or underarms.

Children with NF1 may experience significant problems with their motor skills, such as poor coordination and low muscle tone, and they can be easily fatigued.[4] Throughout life, other common symptoms include vision loss due to tumours on the optic nerve; thinning of bones, which fail to heal if broken; and softening and curving of bones, including scoliosis, a curvature of the spine. Gross and fine motor skills may be impacted, affecting coordination, balance, manual dexterity, reaction times and handwriting.

A genetic test is required to confirm the diagnosis.

But it isn't the physical difficulties that tell us why this gene is so important for understanding the human potential for hugely effective knowledge systems. It is the impact on thinking, on cognition.

The all-important cognitive impacts of NF1

We needed to assess the evolutionary role in knowledge systems that the gene offers us. The clues come from the learning difficulties often faced by those with the NF1 disorder.

Early in her research into NF1, Andrea contacted Dr Vincent Riccardi, known as Vic.[5] When I became involved, the three of us worked on the evolutionary impact and link to knowledge systems. Among many roles, Vic is now Director of The Neurofibromatosis Institute, based in California.

Over a very long career, Vic has probably seen more people with NF1 than almost anyone else in the world. And he is the reason there is a '1' after the 'NF'. There are actually three different neurofibromatosis disorders, which all cause tumours to form on nerve tissues: neurofibromatosis types 1 (NF1) and 2 (NF2), and (equally difficult to say) Schwannomatosis. To add to the confusion, NF1 was formerly known as 'Von Recklinghausen disease', after the researcher who first documented the disorder.

Vic demystified the similar sets of symptoms, recognising that they actually represented three distinct disorders. Eventually, Vic's classification system was confirmed genetically, with separate genes identified as the origin for each syndrome.

It was Vic who named the NF1 protein 'neurofibromin'. Looking back on that decision in the light of many decades of experience, Vic now says that he would have named the protein 'attentin' instead of 'neurofibromin', because he considers that the most common symptoms of the disorder are learning difficulties due to challenges with attention span and spatial reasoning. Both these traits are critical to the argument I'll be presenting in this book.

Vic had always considered that it was the learning challenges for people with the NF1 disorder, not the tumours, that might answer the question about why this gene had survived when, logically, natural selection should have wiped it out. About 20 per cent of children with NF1 will eventually face significant physical challenges. Over 80 per cent of the children with NF1 face moderate to severe impairment in one or more areas of cognitive functioning.[6]

One convincing argument that Andrea and Vic presented to me in those early days was a discovery that Vic and colleagues published in 2018.[7] People with the NF1 disorder are highly likely to suffer from amusia, an inability to hear differences in musical tones. I didn't know what amusia was when the topic was first raised, but on reading Vic's paper, and following up more widely on its effects, I was astounded not only by its existence, but also by the implications for our research. All Indigenous cultures rely on music as a primary knowledge technology. Without it, I can't envisage any of their performance-based knowledge systems being as incredibly effective as they are.

Amusia causes tone and beat deafness

Vic was instrumental—an appropriate word—in the identification of the struggles that those with the NF1 disorder have with musicality. People with amusia cannot recognise a familiar tune unless they hear the words, nor can they distinguish singing out of tune, either in their own crooning or in the off-key melodies of others. This is sometimes referred to as 'tone deafness', but is a far more significant disability than that term suggests (which is more appropriate for people like me who are just really bad singers). Even more critically for our research, people with amusia may also struggle with rhythms, sometimes referred to as 'beat deafness'.

A study involving fraternal twins who had the same formal music training demonstrated that amusia is not due to insufficient exposure to music in childhood.[8] The same study noted that about 50 per cent of the cases also demonstrated a deficit in their perception of rhythm.

With colleagues, Vic described the poor musical aptitude of those with NF1 as 'an inability to sing in rhythm or in tune, difficulty in playing or learning an instrument, and trouble dancing or clapping in synchrony with a song'.[9] Their results demonstrated a highly significant statistical correlation between amusia and NF1: 'the risk of amusia was 42 times greater in patients with NF1 than in healthy controls. More specifically, 67 per cent of the NF1 patients included in this study presented with amusia, while only 4.5 per cent of controls displayed the condition.'[10]

Most reports quote rates of amusia in the general population in the 4 to 5 per cent range. However, Isabelle Peretz and Dominique Vuvan conducted a large-scale sample in 2017, with 20,000 participants drawn from the general public. They reported that 'the prevalence of congenital amusia is only 1.5 per cent, with slightly more females than males, unlike other developmental disorders where males often predominate'.[11]

Given that the rate in the general population is 1.5 per cent, then a 67 per cent rate in those with the incredibly variable NF1 disorder is even more significant. It is important to note that the 1.5 per cent with amusia do not all have NF1. There are certainly other genes involved, but no other gene shows such a strong correlation. If amusia alone was the reason we designated the NF1 gene as being the key to knowledge systems, then we'd be justifiably open to harsh criticism. But it is the combination of a large number of skills, detailed in Chapter 5, that supply the full set of indicators.

If the corruption of one NF1 gene has such an impact on musicality, this tells us that the fully functioning gene gives us something very valuable for our musical skills in the first place. The role of music, both in tone and rhythm, is so critical to Indigenous knowledge systems, which we know date back tens of thousands of years, that I had no doubt that this was a valuable clue as to why this fragile gene survived and flourished throughout evolution.

A number of studies compare their search for the genetic basis of amusia to FOXP2, the language gene. This was very valuable, because it is the research methodology used to link FOXP2 to language that became the model for our formal research project.[12] Peretz and Vuvan wrote:

> The search for the genetic correlates of musicality has gained interest, but it still lags behind research in other cognitive domains such as language. In language, the study of speech disorders has been instrumental in the identification of one of the underlying genes, namely *FOXP2*. Similarly, the study of specific musical disorders, referred to as congenital amusia, an umbrella term for lifelong musical disabilities, may provide new entry points for deciphering the key neurobiological pathways for music.[13]

Oliver Sacks, in his wonderful book *Musicophilia* (2011), presents many fascinating case studies of people with amusia.[14] He considers musicality to be innate in virtually all of us. He asks if music offers a universal human potential in the same way as language. To this question I would answer 'yes', and I would add that the reason can be seen in the way Indigenous cultures sing their knowledge and thus make it wonderfully memorable.

Sacks also considers that humans do not seem to have any innate preference for a particular type of music, but that the indispensable elements are discrete tones and rhythmic organisation. He describes extreme cases of amusia, people for whom a music concert becomes an endurance feat as they are immersed in unintelligible noise.

Sacks explains that people with amusia cannot tell which is the higher note when two notes are played that are close together. While reading his book, I was thinking about the language gene and the knowledge gene, our new nickname for NF1, as separate entities that just work really well together. Sacks describes the confusion he feels when noting that people with amusia can be virtually normal in their speech skills while profoundly disabled musically. He asks if the two can be so totally different, given that speech involves tonal and rhythmic changes which appear to have a musical basis. Maybe this genetic distinction offers a pathway to exploring his question.

The apparent link of music to speech is because of a trait called prosody, which seems to be a separate skill, also impacted by NF1. In linguistics, prosody is the difference between speaking in a monotone compared with adding expression by varying pitch, loudness, duration, stress and rhythms. People exaggerate the prosody in their speech when talking to young children because of the impact dynamic expression has on comprehension and engagement. Those with the NF1 disorder, most of whom suffer musical challenges, often don't perceive prosody in speech. That suggests that our fully functioning NF1 gene also helps us with prosody.

Spatial skills are paramount

Vic noticed that many of his young NF1 patients struggled with spatial reasoning. If given a maze to work out, they would have a harder time with it than the average child. Considering the difficulty they had working out their spatial orientation, it was easy for them to unwittingly turn around and head in the wrong direction. From Vic's observations and the literature on NF1, it is clear that significant problems with visuospatial skills are common.[15]

Visuospatial ability is about being able to manipulate abstract visual images. These essential skills include the ability to mentally rotate objects and to recognise three-dimensional objects from different directions. Your visuospatial abilities enable you to mentally place yourself in space and see your surroundings from different perspectives. Importantly for the art that is critical to non-literate knowledge systems, visuo-spatial skills equip us with the ability to imagine being in an environment and to represent that environment accurately. These skills are closely related to navigation, a perceived sense of direction, wayfinding, tracing a route from a map and pointing in the correct direction. Your visuospatial working memory is constantly helping you to retain and process spatial information.[16]

Visuospatial skills are central to many aspects of Indigenous knowledge systems, including creating art (such as figurines) and portable memory devices, painting on many different surfaces, navigating and using the landscape as a complex memory palace. So critical are visuospatial skills that a huge proportion of the memory technologies used in non-literate cultures could be impacted by a mutation to our NF1 gene.

This is a strong indicator of the potential granted by this gene in the first place. One study concluded that 'visuospatial performance represents a strong predictor of NF-1 diagnosis, finding that children with NF-1 had a particular pattern of performance on visuospatial measures that could correctly identify 90 per cent of over 100 children and adolescents with NF-1'.[17]

In combination with music, the implications of spatial skills being so strongly impacted left no doubt in my mind that a fully functioning version of this gene was essential for the evolution of human knowledge systems.

There are further cognitive clues from the NF1 disorder

The cognitive impacts Vic noticed as the most common among his NF1 patients were attention challenges. The prevalence of attention deficit hyperactivity disorder (ADHD) is high. Difficulty sustaining attention has been identified as an issue in about 63 per cent of children with NF1, with around 35 per cent of them fulfilling the diagnostic criteria for ADHD, compared with 2.8 per cent in the general population. Usually, males dominate the ADHD diagnoses in a ratio of about 3:1. Those with NF1, however, showed similar frequencies in both males and females.[18]

Hyperactivity alone was very rare with NF1. Vic's experience agreed with the many academic reports indicating that ADHD in those with NF1 has a specific profile. Difficulties with attention were reported as far more common than the hyperactivity aspect of ADHD. Children with NF1 have difficulties forming friendships and so are often rejected by their peers, but this problem seems to be closely linked with their ADHD, rather than due to NF1 alone.[19]

In over 60 per cent of NF1 children, and on into adulthood, autistic traits and autistic spectrum disorder were reported. The statistics vary between studies but are always around that level. A full diagnosis of autism for NF1 children ranges between 11 and 26 per cent, which is significantly higher than the 1–4 per cent prevalence reported in the general population. Researchers are particularly interested in this data. 'As a single-gene condition with high autism penetrance, NF1 presents a valuable genetic model for advancing our understanding of the neurobiological mechanisms of autism,' writes one group of researchers.[20]

The autistic traits vary greatly, but tend to manifest primarily as social communication issues. NF1 children often struggle with making social overtures or giving appropriate responses in what they say or even just by smiling. However, restricted and repetitive behaviours are also often present, displaying typical autistic traits such as a dislike of even minor changes along with a limited range of interests. Some reports talk about the presence of rituals and compulsions, but the research varies on how commonly these are observed.

These autistic traits threw up a surprising finding, as I'll explain in Chapter 5. In a contemporary society, these behaviours offer challenges, but that wasn't necessarily the case in oral cultures in the past. It was only when we did the formal assessment of the full skill set in terms of the way people learn in non-literate cultures that we appreciated this conundrum. It led me down a research sidetrack into neurodiversities, as will be described in Chapter 6.

Many of the studies refer to the impact of the NF1 disorder on executive function, which includes organisation or following a sequence. People with NF1 may struggle with planning and working towards future goals. Dealing with abstract concepts can also be a challenge, so solving problems independently can pose issues, as might self-monitoring and impulse control.[21]

Vic reported that he was constantly engaging with speech issues—many of those with NF1 need to be referred for speech therapy to be clearly understood. This may be related to an auditory processing disorder, which is a condition that affects an individual's ability to understand speech.[22] Problems with reading may be related to this.[23]

Memory is at the heart of my work

Most studies indicate that the intelligence of those with NF1 is usually within a normal range. There also seems to be no impact on their verbal or visual memory. This is hugely important. I am going to be talking a lot about memory in future chapters, and I will be distinguishing between natural and trained memory. It seems that NF1 does not have any impact on natural memory, but the ability to train memories will become critical when I show how important music and visuo-spatial skills, among others, are to memory in non-literate cultures.

The challenges faced by those with NF1 provide an idea of the cognitive skills that were foremost in Andrea's thinking when she started exploring the genetics and evolution behind this potentially tragic genetic disorder. Along with Vic, she kept returning to the question of how such a gene could have been so successful in its evolutionary journey as to end up in every one of us.

When Andrea first came across my book *The Memory Code*, she noticed that the long list of traits affected by the NF1 disorder were the very traits that I reported had served humans so well over their very long history before writing was invented.

What could possibly be so valuable that evolution didn't just wipe the initial mutation out? It turns out that the Director of the Murdoch Children's Research Institute in Melbourne and one of the leading scientists in the world specialising in the cognitive aspects of NF1 has been discussing this question with Vic over decades.

The answer lies in our ability to use story, art and music to enhance our learning skills, and in our intense connection to the details within our physical environment. We can use these skills not only to store information and increase our knowledge but also, even more importantly, to explore new ideas. They help us see creative solutions to the way things are, and envisage the way they could be.

CHAPTER 2

Illuminating an ancient knowledge gene

Most of us have two fully functioning copies of NF1 in our genome. Yet this gene is easily broken, leading to a debilitating syndrome. How could this possibly be valuable for increasing your chance of passing on your genes? How could such a gene be favoured for natural selection?

NF1 is considered a high-mutation-rate gene and one of the largest known genes in humans. Studies of individuals with NF1 have identified over 3000 different mutations causing the disorder.[1] The huge range of possible mutations explains why the symptoms vary so much between individuals, yet the impacts on music, art and other knowledge-enhancing abilities used by non-literate cultures are remarkably strongly represented across the whole cohort.

One of the most common causes of the disorder is a 'point

mutation'—that is, a single letter has been changed in the genetic code. A point mutation can corrupt the messages that encode the resulting protein. Other mutations arise from an insertion into, or deletion out of, the sequence of the huge number of bases in the gene. This will throw off the sequence and impact the coding of the protein neurofibromin. Without the correct coding, neurofibromin's tumour-suppressing function can diminish. There's a lot more that can also go wrong, because this gene does far more than just suppress uncontrolled cell growth, but it is the cognitive impacts that tell the story about knowledge. Those changes were initiated when we separated from our nearest cousins, the chimpanzees.

We learn a lot from the chimpanzees

NF1 is a very ancient gene found in many species. Genetically different versions of a gene are known as orthologues. Although slightly different, these genes share the same name because they perform some similar functions in different organisms, and are similar in many key respects. Variations of the NF1 gene are found in an enormous variety of species, from amoebae and fruit flies to mice and primates. The gene appears to have retained its high mutation rate for close to 1 billion years—that is, at least since we diverged from the social amoeba.[2]

But it is the difference from the chimpanzee gene that tells us most about what the human version offers us in terms of knowledge, and how critical this gene is in separating our behaviour from that of all other living species.

We share 98.8 per cent of our coding DNA with the chimpanzee (*Pan troglodytes*). Despite this, chimpanzees haven't adapted to ecosystems all over the world. They haven't developed art, music or complex knowledge systems, and they haven't developed the language skills you depend on to read this book.

Within that 1.2 per cent difference, there are many genes playing a part in all the physical and cognitive aspects that differentiate us from chimps. In terms of learning and storing information, two of them have a very special role: the knowledge gene and the language gene, NF1 and FOXP2.

Andrea asked herself why she had never heard of the tumours so typical of NF1 being recorded in any of the ape species, including the chimps. She consulted with primate specialists. They described cases of leprosy in chimpanzees that produced startlingly similar skin lesions to those in humans, but they were not the same as the neurofibromas known from NF1.

This was a surprise, as it was accepted wisdom that the protein produced by the NF1 gene, neurofibromin, was completely identical to that found in chimpanzees and other apes. One study concluded: 'Human neurofibromin is 100% identical to its orthologues in chimpanzee (*Pan troglodytes*), in gorilla (*Gorilla gorilla*), in orangutan (*Pongo abelii*), in gibbon (*Nomascus leucogenys*), and in marmoset (*Callithrix jacchus*).'[3] There was no logical reason why the neurofibromas wouldn't afflict these species as well.

Andrea makes a major discovery

Early in 2018, Andrea sought the source of this accepted wisdom, but found a lack of evidence for the claim that the NF1 gene was completely identical to that of apes.[4] She found that the cited research had compared only a small portion of the human protein with that of the chimpanzees and other apes.

Using the US National Institutes of Health's genetic sequence database, GenBank,[5] she discovered that there was a single amino acid difference between humans and chimpanzees, near the tail end of the protein. Andrea repeated the search in several ways to ensure that it wasn't just one individual chimp that differed from humans. She also checked gorillas and orangutans, confirming that they shared the chimp version.

A single-letter mutation had caused the distinct human version. If there is a change at a single place in the gene, as Andrea discovered in NF1, then the new gene is referred to as a different allele. Andrea checked other sources, including the 1000 Genomes Project database, an international research project that stores a great deal of human genetic materials, but could not find the chimpanzee NF1 allele in any of the genomes of modern humans.[6]

Contrary to the academic claims, Andrea had shown that the human allele of NF1 is unique to our species.

When did the human allele of NF1 emerge?

There is much debate in scientific circles about when humans diverged from apes, and specifically from the lineage of the chimpanzee. Recent articles put that date at somewhere between about 9.3 million and about 6.5 million years ago.[7] For our purposes, it is just a very long time ago.

What we do know is that at some point after the human lineage split from that of the chimps, one individual belonging to one of our ancestral species, somewhere, somehow, was born with a mutated NF1 gene. Evolution should have wiped out this new mutation quickly. But it didn't. Not only did it leave this mutation hanging around, but it also gradually spread it right through every hominin population for which we have DNA: Neanderthals, Denisovans and us, the modern humans.[8] Denisovans are an extinct archaic human who are only known from a few physical remains from across Asia. Most of what is known of them comes from their DNA.

Andrea spent a week comparing the NF1 gene sequences of modern humans, Neanderthals and Denisovans. She was able to make comparisons through databases such as the Max Planck Institute's Ancient Genomes Browser and the 1000 Genomes Project database.[9]

Neanderthals have been known as *Homo neanderthalensis* for a long time. We now know they interbred with modern humans and produced fertile young. That's why you almost certainly have a trace of Neanderthal heritage in your genome. You are *Homo sapiens sapiens*. It is now common to refer to Neanderthals as *Homo sapiens neanderthalensis*, recognising that we are not really separate species. Denisovans have not yet been assigned a scientific name, mainly because too little is known about them physically. As some modern human populations have traces of Denisovan DNA, it is likely they will also be classified as a subspecies of *Homo sapiens*.

Andrea found the human NF1 allele in all the ancient Neanderthal and Denisovan individuals for which she could access the DNA. There was absolutely no difference in the coding sequences of these ancient NF1 genes. Given that the last common ancestor for humans and Neanderthals is now estimated to have been around 500,000 years ago, and that we share this allele, the likely emergence of the initial mutation was before that date.[10]

A unique human allele started with one little baby

One day, a hugely long time ago, there must have been a mutation in a single egg or sperm of one of the species on our ancestral tree. Just one time, one event. Mutations are happening all the time, and most pass unnoticed. This one was probably not noticed at the time, but it was a massively significant moment when the first offspring was born with one of the new human NF1 alleles in its genome. Evolution then took over.

Evolution happens at the population level, not at the individual level. There are four forces of evolution currently recognised. The first is mutation, which was obviously critical in initiating the existence of the human version of NF1. The second is known as 'gene flow'. It occurs when a significant number of individuals move into or out of a population, changing the frequency of the alleles within the population. If this was a significant force for NF1, it was a very long time ago and likely not something we can ever know. The third force, genetic drift, is a random change in allele frequencies in a small population, and unlikely to be of significance here. It is the fourth, natural selection, which we'll be talking about in this discussion of the evolution of the knowledge gene.

Natural selection occurs when individuals with a certain genetic make-up are more likely than others to survive and reproduce. Over time, more individuals with the newly unique NF1 allele must have passed it on to the next generation than those without it. And so on, generation after generation, until everyone had it. This again raises the question: why did this new human NF1 allele completely replace the ancestral chimpanzee type? The new NF1 allele must have been favoured because it provided a significant advantage for the population as a whole through the individuals who inherited it.

There was clearly a cost associated with this new NF1 allele in the form of the disorder, but that needed to be weighed against the benefit for the population as a whole. Evolution performs that balancing act constantly.

A wonderfully clear example of a gene mutation that spread through most of the relevant population is that which causes sickle-cell anaemia. Imagining an anthropomorphised version of evolution, the sickle-cell anaemia example provided me with a valuable insight into what might have happened with NF1.

Natural selection as a genetic cost-benefit analysis

Unfortunately for some, evolution does not exist to benefit individuals. It even appears to sacrifice the extremes for the sake of the majority. Natural selection is not always a kindly player, as those with sickle-cell anaemia will affirm.

The sickle-cell anaemia mutation has persisted for at least 5000 years. It is often used as a case study for describing natural selection.

Sickle-cell anaemia affects millions of people worldwide. It is a recessive genetic disorder that is most common in Africa, but is also found in countries around the Mediterranean and eastwards as far as India. In the Americas, it mostly affects people with African heritage. One in twelve African Americans is estimated to carry at least one copy of the allele.[11]

The sickle-cell anaemia allele leads to tragic consequences when a person inherits two copies of the mutated gene, including the likelihood of a significantly reduced life span.

Normal red blood cells carry oxygen around the bloodstream to cells throughout the body. Sickle-cell anaemia affects the haemoglobin protein in red blood cells. The mutated gene changes the red blood cells from their normally round shape with a depression on either side to a stiff, sickle-like crescent shape when stressed, due to low oxygen or dehydration. Because of their elongated shape and inflexibility, the deformed cells tend to form clumps in the blood vessels. This causes episodes of extreme pain because the clumps inhibit blood flow, leading to problems in the oxygen-deprived tissues. As the sickle cells often last only twenty days (compared with 120 days for normal cells), the overall shortage of blood cells can cause low-iron anaemia, leading to extreme fatigue, shortness of breath and hindrances to a child's growth and development.

Why would evolution maintain the sickle-cell anaemia allele in some populations at such high levels?

The answer was found when scientists noticed that countries with high rates of sickle-cell disease also had high rates of the potentially fatal disease malaria. It turned out that people who carry the sickle-cell allele are far less likely to experience a severe case of malaria. It is not those with two copies of the recessive gene who benefit most, given their shortened expected life span and episodes of pain. People who carry one normal allele and one sickle-cell allele have both the sickle and round blood cell types in their bloodstream. They produce enough red blood cells to avoid the symptoms of sickle-cell anaemia, while having enough sickle cells to provide protection from malaria. The much larger number of people with only one copy of the recessive gene benefit, with a cost to the smaller number who carry two copies.

The evolutionary link between sickle-cell diseases and malaria was postulated in 1949 and confirmed in 1954.[12] Evolution has allowed a cruel burden to be endured by a small proportion of people in these regions in order to enable a larger proportion of the population to benefit.

Cost-benefit analyses like this explain why the unique human NF1 allele managed to survive despite its potentially tragic consequences. The ability to store knowledge really is that important.

The appearance of the NF1 allele

To analyse the value of the human NF1 allele in terms of the evolution of knowledge systems, it was important to know how long ago the whole population began to have two copies of this allele.

At the time Andrea was hunting through the genetic databases, the oldest hominin sequence in the Ancient Genomes Browser was 126,000 years old. That Neanderthal carried the human-type allele on both copies of chromosome 17. Andrea realised that this narrowed the origin of the new allele to sometime after our split from the chimp/human common ancestor, at least 6 million years ago, but before the birth of this Neanderthal's parents.

The only certain conclusion Andrea could make was that the human allele must have emerged before the last 126,000 years, even though evidence from the last common ancestor for humans and Neanderthals suggests the human allele might be over 500,000 years old. Data from the 1000 Genomes Project showed that the chimp allele has never been found in any modern human population.

This means that we can look at human prehistory for the last 126,000 years certain that everyone had the same NF1 allele pair as you carry today. Consequently, this period will be the focus for the rest of the book. I haven't been able to find anything in the archaeological record pointing to art or music or spatial awareness from before that time anyway, possibly because, with artefacts deteriorating, little is available for archaeologists to find.

Cro-Magnon 1 suffered with NF1 around 30,000 years ago

We know that the human NF1 allele has been around for more than 100,000 years, probably many times that span. But what about the NF1 disorder? We can surmise that modern humans have long suffered with severe symptoms of the disorder because of a 30,000-year-old human skeleton known as Cro-Magnon 1 (see Figure 2.1).

In 2018, a paleopathology team from Paris reported their NF1 diagnosis for a skeleton with a strange forehead profile.[13] Cro-Magnon 1 had been discovered in 1868 in a rock shelter called Abri de Cro-Magnon in Les Eyzies, France. The male, the first of several ancient skeletons found in the rock shelter, was estimated to have been around 40 at the time of his death. He had a large, rough indentation on his forehead, as well as a number of lesions in various other bones.

Using modern technologies such as CT scans to examine both the surface and the interior of the bones, the French

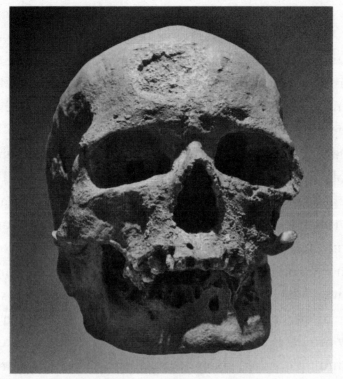

FIGURE 2.1 Cro-Magnon 1 skull, one of the individuals from the Cro-Magnon rock shelter, showcase moulding from the Musée de l'Homme, Paris.
PHOTO: 120 (CREATIVE COMMONS LICENCE CC BY-SA 3.0).

team determined that the most likely cause of Cro-Magnon 1's deformities was neurofibromatosis type 1. The team also attempted ancient DNA testing, but Cro-Magnon 1's bones have not yet yielded any usable DNA.

It appears that the human allele was causing tumours in ancient populations in the same way it does today.

Another new word for me: NF1 is a *supergene*

As I was learning about genes, Vic and Andrea were talking about 'supergenes'. The previous chapter discussed the large array of physical and cognitive impacts caused by a mutation of the NF1 gene in the egg or sperm that created a modern child. Yet because it is referred to by that single name, NF1, you can get the impression that it is only a single gene. Every gene is far more complex than the trait or traits that it is best known for. And NF1 is more complex than most because it is considered to be a supergene.[14]

Many people incorrectly assume that one gene typically controls one trait, and that each trait is determined by a single gene. These one-gene one-trait assumptions are almost never true.

With regard to eye colour, it is often said that 'green and brown eye colour genes are dominant, and blue eye genes are recessive'. This is vastly oversimplified. Multiple genes contribute to eye colour, and most, if not all, are involved in other bodily processes as well. So the average gene works in concert with multiple other genes located across multiple chromosomes. In this way, they influence a variety of physical or behavioural traits. Having genes located on different chromosomes means the inheritance of the related traits can vary greatly with each offspring. That diversity is massively valuable to the population as a whole.

A supergene is a bit different. It is a collection of distinct genes that have been combined into a single genome region so that they are almost always inherited together. They effectively become a single gene with multiple functional areas. Vic Riccardi has long argued that NF1 is a supergene. With ten different identified functional regions and three other genes embedded within, it is easy to see why.

In the human NF1 allele, only 9000 of the bases are needed to actually encode the protein neurofibromin. These crucial segments of code are interspersed across a massive 300,000–letter sequence of mostly gibberish. This is much longer than most genes. The great length contributes to making NF1 a supergene, but also explains its susceptibility to being broken.

In general, the most ancient genes are the biggest because they've picked up a significant proportion of non-coding DNA along the way. NF1 is enormous. Mutations causing the NF1 disorder have been found across the entire length of the massive gene. It is therefore not surprising that the disorder manifests in so many different ways, impacting on so many different physical and cognitive domains.

What really matters is that the research into NF1 revealed that art and music, story and our visuospatial skills have acted in concert in the long history of human knowledge systems. We should look at the entire complex package, rather than at any of the skills in isolation.

Learning from FOXP2, our language gene

Andrea had hypothesised that the benefits of NF1 lay in the improved ability to store knowledge through memory systems. Now we needed to test that hypothesis, and we were fortunate to have an excellent precedent to follow.

Way back in 2002, news of a language gene first made the headlines.[15] The gene FOXP2 is hugely significant in enabling human language. Working together, language and memory give us a potential no other species can claim.

The role of FOXP2 in our ability to use complex spoken language was first identified through the study of a family in Great Britain, referred to in the medical literature as the

KE family. Many members of the family had an inherited severe speech disorder called developmental verbal dyspraxia, which affects both the production of speech and language comprehension. Tests of their DNA revealed a mutation in the FOXP2 gene that was shared by all family members who exhibited the language challenges, but not by the members who didn't. Further study revealed that the fully functional human FOXP2 gene is different from that in chimpanzees.

The chimpanzee FOXP2 gene sequence is shared with other apes, confirming that, as with NF1, the differences arose among our ancestral species sometime following the split from the joint chimp/human ancestral population. Variations of the FOXP2 gene are found across vertebrate species, with FOXP2 gene mutations documented to affect communication in other animals. For example, the ability of songbirds to produce the vocalisations typical of their population is impacted if their FOXP2 gene is mutated.

FOXP2 really does appear to be a communication gene for all animals with it, and the uniquely human FOXP2 allele appears to play an important role in both the complex sounds that humans can use in speech and in our ability to make sense of them.

How fast can a new mutation spread through a population?

As we saw with Covid-19, viruses can replicate, spread and mutate at lightning speed. But the changes I'm talking about here happened much more slowly. Virus reproductive cycles are measured in hours; human ancestral populations, like us, will have had generational cycles measured in decades.

At some point, DNA in the sperm or egg cell of one of our ancestors developed a FOXP2 mutation that was then passed on to their offspring. A baby would have had one of the new human FOXP2 alleles and one of the old ape-like variants. They may have grown up with a slight advantage, because their genome had a single copy of the new allele. I like to imagine that very first individual, who can have had no idea of the impact their life would have on everyone reading this book now. If they had not bred, then we might not be talking or reading. But, much as I enjoy speculating on alternative scenarios, I should return to what probably did happen.

The children of that first FOXP2 hybrid each had a 50 per cent chance of inheriting the new allele. And so it went on, until these FOXP2 hybrids were numerous enough that two of them met and bred and produced a child with two copies of the new allele. As more children with two alleles appeared, they probably had some kind of communication advantage. Certainly, as more and more generations of doubly advantaged offspring existed in the population, the potential of communication would have become apparent. The ancient Neanderthal and modern humans carried two fully functional copies of the human FOXP2 allele, underpinning a full potential for language.

At some stage, the same process must have happened for NF1. When and where, we simply don't know. We are talking about timespans measured in thousands of years. What we do know is that, eventually, individuals with two copies of the not-so-new FOXP2 allele met individuals with two copies of the now-established NF1 human allele, and they bred.

Both genes were so valuable for human populations that they became what geneticists call 'fixed'—that is, they are found in every member of human populations for as far back as we can know with current technology.

In humans and across species, FOXP2 does not demonstrate a mutation rate nearly as high as that of NF1. But FOXP2, having been studied for decades already, provided us with an ideal model for approaching our investigation of NF1.

This is not a single-gene theory. The claim that NF1 is our knowledge gene is justified because it contributes so significantly to our knowledge systems in a way that no other known gene does. There is no doubt that other genes will contribute, but without the connected suite of skills provided by NF1, we would not have the ability to store information so essential for human evolution. Looking at evolution through the lens of the NF1 gene allows us to explore human knowledge systems in a way that has not been possible before.

It is also clear that we need FOXP2. They work together. I am now convinced that the skills described in this book could not have arisen without NF1. We could not have become the thinking, learning, knowing creatures that we are if that ancient mutation hadn't taken place.

I join the team

For a few years, Andrea had been asking why evolution had left NF1 in the gene pool despite the cost to so many individuals. Her head was also full of all the complexities of genetics and evolution from her experience in the field. Andrea and Vic had been collaborating for a few years, and I was excited to work with them.

As a team, we did science differently, but not because we planned it that way. Scientific discoveries usually arise from specialists within a single discipline building on existing knowledge. Their work delves deeper into a narrow aspect of their discipline than their colleagues before them. The three of us came from different academic disciplines.

Once we had decided to embark on a formal research project, my next step was to review over a decade of my work and look at it differently. I had written a lot about the memory systems of Indigenous cultures, but now I needed to think of them in terms of the specific cognitive skills employed so that we could evaluate them against the skills impacted when an NF1 gene goes rogue.

CHAPTER 3

A fresh look at Indigenous memory systems

Over a decade before I first learnt that NF1 existed, I explored how Indigenous Elders managed to remember so much pragmatic information essential for their survival, both physically and culturally. I had focused on the way memory systems worked so effectively, but not as much on the specific skills employed by Indigenous knowledge keepers. What I had written resonated strongly with Andrea's understanding of the cognitive skills impacted by NF1—but was that just a superficial match?

I went back to my previous research, including much that had not made it into the book Andrea had read. Immersing myself in this again prepared me for the formal part of our research project, when I would create a checklist of all the skills in Indigenous knowledge systems. This would enable Andrea and Vic to compare them against the skills impacted when the NF1 gene no longer functions properly. There would have to be a significant overlap if we were to claim that the answer to the NF1 evolutionary conundrum lay in my approach to memory.

At this stage, the skills that had most attracted my attention were musicality and spatial abilities. Oral cultures rely on performing knowledge, which makes it incredibly memorable. I had long been intrigued by the way Australian Aboriginal songlines involved vast amounts of knowledge encoded in songs and dances performed right across the landscape. So my first step was to review my ideas of the skills required by people living in a culture dependent on a performance-based knowledge system. I needed to think about the big picture, not just the specific domains that the three of us had already discussed.

Andrea and Vic asked me not to engage with the NF1 literature as presented in the previous chapters, so that my checklist would not be influenced by any more detailed knowledge than the superficial insight I already had.

More similar than different

L.P. Hartley famously wrote in *The Go-Between*: 'The past is a foreign country; they do things differently there.' People in the Neolithic were way in the past, and so very different. In the Palaeolithic, they must have been even more different. And there are Indigenous cultures the world over. They are very different from my Western upbringing and very different from each other, and probably even more different from the cultures that we know only from the archaeological record.

Or are they?

If you focus on skin colour, subsistence, population densities, habitats . . . yes, there are many differences. However, all humans obviously eat, breathe, are born and die. All humans show emotions, care for their children, fall in love, argue and defend. All humans share 99.9 per cent of our coding DNA. And we all have at least one fully functional copy of our human NF1 allele.

When it comes to the cognitive skills associated with knowledge, the similarities across all non-literate cultures become overwhelming. It is these similarities that will answer the question: what is it that NF1 grants us that evolution considered so hugely valuable?

With my background in science, I focus on the knowledge domains that are shared between all cultures: those with the rational knowledge of their environment and everything in it. In what is now seen as an outdated and offensive attitude, mid-20th-century anthropologists treated Indigenous cultures as exotic, and so highlighted all the things that make us different.

But it is our similarities that make us all human, not our differences.

How do they remember so much stuff?

Cultures that do not depend on writing to store information—oral cultures—have a whole swag of methods to make knowledge memorable. They use song, story, dance, mythology and multiple physical devices to memorise everything they need to survive, and even more that they just want to know. Too little credit is given to the intellectual achievements of traditional Indigenous cultures.

Imagine if you were asked to memorise the name of every one of the hundreds of animals you might encounter: large, small and tiny. You will need to know their habitats, behaviour and seasonality, as well as which of them are dangerous and which might be dinner. Add in a good few hundred plants, again with all the details for gathering and using them. It can be fatal to mix up those that are edible and those that are not. Some can be used for medicine and some are superb for binding wounds.

You need to memorise where to find water in a desert, and how to identify a tiny herb in a forest. You'll need to know how to manage the land and navigate hundreds of kilometres around it, day and night, without tracks, let alone roads.

Given that you'll need to travel long distances to gatherings for ceremony, trade and finding partners, it's worth making sure that you don't leave a week late. To maintain a calendar, you need to remember detailed observations of astronomy and seasonal changes. To feed yourself, you need to understand weather conditions and climate, and the various habitats. To make tools, you need to understand geology and where to get the best flint or obsidian, be that by digging or trading.

To live in a society, you also need to memorise laws as well as ethical expectations. Then it gets complicated. You know that nice, neat family tree you have—or that you intend to work out one day? People of Indigenous cultures are known to memorise vast genealogies, and usually all the relationships, not just a neat tree of descendants.[1] Most famous in this regard are the male Griot and female Griotte of West Africa. They will sing their long genealogies and the associated narrative history for hours, even days.[2] Indigenous genealogies might consist of lists of names, which act as memory prompts for further information. In New Guinea, for example, the mythology and the clan songs memorised for the Iatmul are linked to an enormously elaborate totemic system of an unimaginably long list of thousands of multisyllabic personal names.[3] Anthropologist Jan Vansina wrote that Indigenous genealogies 'are among the most complex sources in existence'.[4]

By now you should have the idea that Indigenous teaching involves memorising, understanding and using vast amounts of information. We are not just talking about people who could do well on a quiz show. Not even people who could do brilliantly well on a quiz show. We are talking about people who have trained their memories from childhood to do brilliantly in all areas of general knowledge.

Just imagine what it would be like to store all this information in your memory and keep it accurately for the rest of your life. That is what First Nations Elders have been able to do. Their encyclopaedia is the result of many thousands of generations compiling and testing information and handing on what is important. What is not important is gradually lost from the retelling and from the oral tradition. Oral tradition is dynamic, forever changing, forever being updated.

Let's be practical. You can't constantly retest every plant in the environment to find out what is poisonous, what is edible and what can be used for building or art. You would have half the population doing nothing else but testing, and some would die, literally, for the knowledge. You can't relearn everything about animal behaviour or astronomy, or develop new laws and ethics every generation. Some information has to be retained accurately. The experiences of others in the past prepare today's people for rare eventualities, such as extreme droughts, unusual illnesses or injury, or the failure of animal migrations.

But beyond identification, you need to know how to deal with the plant or animal. For example, a number of the many Australian Aboriginal cultures depend on cycad seeds (*Macrozamia moreii*) as an invaluable food source, especially for large gatherings at ceremonies. There's only one problem: fresh cycad seeds are deadly. Detailed methods need to be memorised and applied in order to treat the seeds over a number of days to make them edible. This knowledge was discovered long in the past and carefully handed down, very accurately, to generation after generation.

And there are hundreds of local plants in every habitat. Some are of no use, but Elders need to know them all to avoid constantly retesting those that have previously been found to lack practical application.

Indigenous cultures tend to maintain their knowledge as an integrated whole, not in nice little silos as we do in Western universities. As Indigenous American Dakota astronomer Jim Rock explains:

An indigenous world view or philosophy is interdisciplinary. It doesn't mean that 'everything is thrown together in some random way'. That's unfortunately the outside perception. But we're not afraid to see patterns of patterns. We're not afraid to see the metapatterns and the patterns of connections. Being an astronomer, that's how we see those stars, constellational patterns. So I call that [interdisciplinary view] a 'constellation of thoughts'. There's a weaving, so it's interdisciplinary, like the music piece or the storytelling piece may have multiple purposes, objectives and multiple seeds that are planted in our soul that grow and bear fruit over time. There's the mathematics that's in the music and the ethics and values—so it's interdisciplinary.[5]

For a Western ethnographer to explore any of the genres of knowledge, they need the knowledge keepers of the culture to work with an array of non-Indigenous specialists in each of the many fields, along with linguists capable of translating nuances accurately.

Take, for example, the North American Navajo and their knowledge of insects. It took at least two Navajo Elders working with non-Indigenous entomologists and very competent linguists to compile the list of over 700 insects stored in the memory of the Navajo knowledge keepers.[6] Only one of these insects, the cicada, is used for food. Ten more are known because they are bothersome, such as gnats and fleas. The entire entomology is known because, like all intelligent humans, the Navajo value knowledge. The Navajo insect classification is included in stories intertwined with ethical, philosophical and mythological themes.

A similar team was needed for research with the Hanunóo in the Philippines to identify the 1625 Hanunóo plant types, far more than Western botanists knew in the middle of the last century, when the research took place. Nearly half of the plants were deemed edible, and another quarter used purely for their medicinal properties.

Our goal here is to understand how oral cultures retain so much information. That knowledge, and the methods used to retain it, are valuable clues to why the NF1 gene ended up in all of us.

I became confused by some troublesome terms

When I started down the unfamiliar track of Indigenous knowledge systems, the use of language in older anthropological and archaeological reports caused me confusion. Often the terms 'oral tradition' and 'oral history' were used interchangeably. A huge proportion of Indigenous knowledge held within oral tradition is not about history. So oral history is a subset of oral tradition. Oral tradition includes all those pragmatic topics talked about in the previous paragraphs. History is incredibly important for some cultures, such as the Māori, whose knowledge keepers can recite 800-year genealogies dating from the time their ancestors first settled New Zealand. For the many Australian Aboriginal Nations, with a continuous culture dating back over 65,000 years, chronological history is far less important than the vast array of knowledge they consider essential.

In the first book I opened about Stonehenge, the archaeologist author referred to the Neolithic Brits as 'illiterate'. That is hugely misleading. Cultures without writing are often referred to as illiterate or non-literate. That's describing them in terms of what they don't have. I could describe you as non-avian because you can't fly, or non-feline because you're not a cat. Maybe you are non-white or non-black or non-Tanzanian. Describing you in terms of what you are not is fairly pointless.

'Illiterate' people are those who cannot read while living in a society that uses writing. 'Non-literate' or 'oral' cultures are those that have no dependence upon writing; they will be the focus for the next few chapters. It is their memory methods that give us the greatest insight into what might have been going on in the distant past.

Cultures without any contact with writing at all are better referred to as 'primary oral cultures'. It is the memory technologies of these cultures that are referred to as 'primary orality'.[7] Later, I will be using the term 'pre-literate' to refer to those cultures that went on to develop writing themselves, such as the Neolithic Chinese.

It is also important to be careful when using terms such as 'spirituality' and 'ritual' and 'religion'. These are too often thrown around almost interchangeably, or interpreted from a Western sensitivity. 'Spirituality' has a different meaning for each Indigenous culture. You cannot assume that they all have gods. Many First Nations people talk in terms of ancestors, and the term I most often hear is 'ancestors whose stories we tell'. 'Ritual' means a repeated act; I found that I could not assume that it meant any more than that. Rituals don't always have a spiritual or religious aspect, although often they do.

The assumption that prehistoric people had gods whom they worshipped is questionable. The term 'gods' and 'worship' are not ones I hear from First Nations people, but are common in interpretations by those writing about them. In his extensive collection of Central Australian Aboriginal songs, Australian anthropologist and linguist T.G.H. Strehlow noted that there was nothing he considered the equivalent of prayers to the spirits or to the totemic ancestors contained in the songs he was recording.[8]

I am constantly annoyed by the portrayal of Indigenous people as if they live in a permanent fog of superstition. It seems that even some academics believe that logical, rational and scientific thinking only started with the Greeks and literacy. The arrogance of literate colonisers often led to the assumption that anything they didn't recognise was just the silly behaviour of scientifically ignorant natives. Rituals they didn't understand became 'magic', rejecting the idea that these performances actually had an effect on the outcome of the hunt.

I have talked with Indigenous people in both North America and Australia about the term 'magic'. They told me that the rituals performed before a hunt not only increased the kill, but also reduced the risk for the hunters. The reason? The 'hunting magic' consisted of songs, stories and dances that enacted the strategies for the hunt, the behaviour of the target animals and group tactics. They also bonded the hunting team.

Mythology rules, okay?

The word 'mythology' is another troublesome term in my early reading in the field. The implication is that a myth is a fictional account. Western culture divides narratives into fiction and non-fiction, and never the twain shall intersect. But English simply doesn't have a suitable word for stories that encode knowledge and whose historical veracity is often irrelevant. Many of the stories are clearly historical accounts, while others are obviously metaphorical. Some are neither. All are adapted to the performance nature of the knowledge system and to current political and cultural needs.

So is any particular story true or false? Are the stories about biological forebears or mythological ancestors? The question is the problem. This is a distinction for historians, but usually not for the Indigenous people telling the story. Mythological beings may be biological ancestors who, over time, have become mythologised. They may relate to observations of the sun, stars, an animal or any other origin. It will depend on the specific culture and their priorities. What matters when talking about being human is that, in all cultures, mythology stores a vast amount of pragmatic information, along with spiritual and cultural understandings.

The inadequate English word used as a general term for Australian Aboriginal knowledge systems is 'Dreaming'. It is always preferable to use the Indigenous name for the specific knowledge system and avoid the 'truth' question altogether. For example, *tjukurpa* (pronounced 'chook-ORR-pa') is the complex of knowledge for the Anangu culture of Central Australia. There is no English word that encapsulates their description. In the words of Anangu people:

> *Tjukurpa* stories talk about the beginning of time when ancestral beings first created the world. These stories contain important lessons about the land and how to survive in the desert as well as our rules for appropriate behaviour.
>
> *Tjukurpa* stories are also used like maps. They tell us where important places are, how to travel from one place to another, and where and when we can find water and food.
>
> The stories are not simple. They represent complex explanations of the origins and structure of the universe and the place and behaviour of all elements in it. The

understanding and depth of these stories increases through–
out a person's life.

Tjukurpa is not written down. It is memorised and passed
on to the right people like an inheritance.[9]

I live and work on Dja Dja Wurrung Country, in the south-
east of Australia. The local Elders have asked that I refer to their
stories and the associated performances as 'teachings'—a far
more appropriate term than 'mythology'.

Mythology may emerge over time from historical events.
Belgian anthropologist Jan Vansina described a historical
quarrel between the Indigenous American Hopi and Navajo.[10]
The outcome was a border division of the disputed land.
To record this agreement, the story needed to be repeated.
Slowly, after 80 years, the narration had been enhanced, with
mythological characters acting out the oral history. It now
resembles what would be described by many anthropologists
as mythology, and is retold to formally record the location of
the border. Is this myth? Is it now fiction? Or is it knowledge
that has deliberately been made memorable and reliable by the
addition of mnemonic technologies?

Mythology, as it exists in oral traditions, is a powerful and
invaluable mnemonic device. The stories bring to life vibrant
and active characters, who are often stunningly beautiful or
horrendously hideous, hysterically comic or vulgar in the
extreme. This makes the story extremely memorable because
of the emotional response it generates.

Given the importance of these mythological characters to
encoding critical knowledge, it is not surprising to find that
figurines, masks, paintings and other representations of the

characters dominate the art traditions of Indigenous cultures globally. Art is a knowledge technology.

Kangaroos hop and grizzly bears chomp on unfortunate hikers no matter the colour of your skin. Knowledge from observations of the real world is common to all of us. What isn't common is the way that information is stored, and that is key to understanding the massive advantage the human NF1 allele gave us. All the skills associated with creating, maintaining and imparting mythology are integral to non-literate knowledge systems.

Was I over-interpreting mythology?

Self-doubt ruled my world at the early stages of my research of Indigenous science knowledges. I was becoming more and more astounded at the complexity and accuracy of the natural sciences stored in First Nations stories. Indigenous colleagues explained that my confusion was partly that science isn't isolated in their knowledge system the way it is in the Western tradition. Exacerbating my struggles was the fact that the only stories I had access to were the public stories, as told to children.

Even so, I still feared that I was over-interpreting. A lifetime of immersion in Western science had never even hinted to me that there were alternative ways of knowing. Was I believing the research I wanted to believe, and searching out the research I wanted to find? One academic paper implied that I might gain insight into the answer from the combined work of Tewa writer Alfonso Ortiz and ethnobotanist Richard I. Ford on corn. I was desperate to get copies of their works, but neither were available in my library, nor online. I was also intrigued because my favourite crop from our vegetable garden was corn, but we had never planted anything other than the standard yellow varieties.

I soon acquired Ortiz's book, *The Tewa World* (1969), but it was very hard to track down Ford's paper, 'The color of survival'.[11] After three months, the university librarian called. She had secured an original copy of the journal. It was so musty that I read it outside to reduce the impact of the smell. But with every word, I became more excited. Ford and Ortiz, in combination, showed me how two completely different cultures might store information totally differently, but with the same observations, understanding and outcomes. I was elated.

Planting a monoculture of yellow corn would be a potentially devastating crop for a Pueblo community, should adverse weather conditions prevail even for a single season, as it often does in the south-west of North America. The Pueblo have a variety of colours in their corn, including yellow, red, black and white, each of which is best suited to different environmental conditions. Corn cross-pollinates readily, so without strict guidelines, the Pueblo would soon lose all their pure varieties.

The Indigenous American scholar Ortiz and the Western-trained ethnobotanist Ford both write about the Pueblo rules for planting the various colours of corn, which have managed to keep the varieties pure over hundreds, probably thousands of years. Ortiz tells of the mythologies of the Corn Mothers and the Corn Maidens, each representing different colours, while suggesting we also read Ford on the topic. Ford, who recommends we read Ortiz, explains how these stories encode the way in which the purest kernels are selected for each colour. These are then planted according to a complex set of rules relating to the spacing of the plantings, in both distance and time. Rather than maximising the yield in any given year, the Pueblo ensure that the population will be fed despite the challenges of a harsh, unpredictable climate. Whichever way the knowledge is expressed, the outcome is the same: the corn varieties are kept pure and the Pueblo survive.

The knowledge of plants and related topics is never kept at a superficial level in oral cultures. It took 500 pages to record the traditional medicinal practices of the Matsés peoples of Brazil and Peru to prevent its loss as knowledge holders die.

So do the knowledge holders in oral cultures know that they are using mythology in this way? Of course. Anthropologist John MacDonald quotes Inuit Elder Hubert Amkrualik:

> 'Stars were well known and they were named so that they could be easily identified whenever it was clear. They were used for directional purposes as well as to tell time . . . stars could be remembered by the legends associated with them. The people before us had no writing system so they had legends in order to remember.'[12]

Without writing, storing essential information depends on memory

We all struggle with an unreliable memory, some of us much more than others. Despite common belief, there is no such thing as a photographic memory. If there was, such people would win all memory competitions effortlessly. All memory is flawed to some degree.

Walking around with a smartphone is a very recent thing.

Go back only a few hundred years and paper and pencil weren't easily accessible. Go back a few thousand years and writing wasn't an option. There were no scripts. Our ancestry includes tens of thousands of years of behaviourally modern humans, and hundreds of thousands of years of ancestral hominins. Given the slow rate of evolution, that means your brain is still set up for memory. The pity is that very few of us use our incredible potential to build a vast knowledge bank.

Until the development of writing about 5000 years ago, all people depended on their memories for everything they knew. Even in literate cultures, it is only in the last few hundred years that any other than an educated elite were able to read. It is not surprising that humans developed the most astonishing suite of memory technologies to optimise their ability to record knowledge.

The ancient Greeks talked about both 'natural memory' and 'artificial memory'. An artificial memory is what we would call a trained memory. If you depend on your natural memory alone, you are using only a small portion of your genetically granted memory skills.

Indigenous people the world over survived, ate well and flourished, even as they migrated into new and unknown environments. The many TV shows that dump city-dwellers in the wild to survive as best they can shows how difficult it is without knowledge of which plants and animals they can eat and which will kill them. Foraging communities need to be able to identify weather conditions and seasons to indicate the timing of resources and to know how to find those resources across a broad landscape throughout the entire annual cycle.

To answer the evolutionary question about the human

NF1 allele, we can learn most by concentrating on hunter-gatherer-fisher cultures, those who depended on wild foods, because this lifestyle dominates the vast majority of human history.

I will mostly consider knowledge systems relating to the concrete reality, their scientific knowledge of the world around them. Religious beliefs, cultural expectations and politics vary hugely. Knowledge of the natural world—how people gather food and survive—allows us to compare information from the same data set: the environment. It is this knowledge that enables humans to move across the world, to adapt and to flourish, to learn and to teach, in a way no other species has managed.

If you are reading this book, then it is highly likely that you depend primarily on the written word for your knowledge storage. But, given that humans have lived in oral cultures for the vast majority of their evolutionary existence, our brains have evolved in a way that allows them to efficiently store knowledge without using writing. Today, almost all oral cultures have contact with writing, but we can still gain significant insight from them about how primary oral cultures would have functioned.

So what are these incredible memory technologies? What skills are needed to apply them? They are things you know all about already—you probably just don't use their full potential to help you learn. They are song, story, dance and, as we'll see in the next chapter, connection to place, as well as a whole range of physical devices. Amazingly, our NF1 allele seems to have enhanced both aspects of the memory technologies— oral and physical.

Knowledge is performed through song, story and dance

Which is easier to remember—a page of text you once read or a dynamic performance you once experienced? For almost everyone, performances with music and vibrant characters acting out exciting stories are far more memorable than strings of facts. That is why all oral cultures perform their knowledge. Song, dance, narrative and audience engagement are critical to every Indigenous knowledge system in the world.

The performative nature of Indigenous knowledge systems is one of the reasons traditional knowledge is so difficult to translate into the written word. When linguist Cliff Goddard was documenting the knowledge contained in the songs recording over 100 plants of the Australian Yankunytjatjara people, he commented that 'in written form the stories lack the performance quality that was so much a part of the way they were told'.[13]

Some performances may be small, for just a few people or the local group. Others may be part of huge gatherings. Although infrequent, the ceremonies for Australian First Nations could attract thousands of people, who would travel great distances to attend. From Aboriginal corroborees to Indigenous American powwows, gatherings the world over serve a multiplicity of purposes, including trade, meeting marriage partners and socialising. Here, we will concentrate on the role of these gatherings for maintaining, updating and sharing knowledge.

I mentioned *updating* knowledge. There is no culture on the planet, and there probably never has been, that is stagnant. Not only does new knowledge about animals, plants and the environment often need to be added, but there can be a political motivation as well.

The Senior Indigenous Curator at the National Museum of Australia, Margo Ngawa Neale, describes the performances at Australian Aboriginal gatherings using the metaphor of the human body.[14] She describes the knowledge included in the performances as being like the human skeleton: every body has the same set of bones, in the same structure. Every performance of a given aspect of knowledge is also rigorously protected and kept accurate. But just as we each have different flesh on those bones, every performer adds their own interpretation, their own presentation, and that individuality makes the performances more entertaining and therefore more memorable. Critically, new information can be added; if it is deemed valuable, it may eventually become part of the skeleton. The top performers in a community are highly regarded and valued. Yanyuwa Indigenous Australian Eileen McDinny put it like this: 'Everything got a song, no matter how little, it's in the song—name of plant, birds, animal, country, people, everything got a song.'[15]

Annoying as it may be when a song gets stuck in your head and repeats over and over and over, that phenomenon is an invaluable mnemonic technology. Try describing a bird call in writing. If you check your local field guide, you'll discover that writing down bird calls doesn't work very well. Now try mimicking bird calls in song, or the way an animal behaves in dance. Singing and performing animal behaviour is a far more effective method of storing this kind of information. Coupled with the rhythm, rhyme and tunes of singing is the urge to move, to dance. Indigenous knowledge systems the world over are 'embodied'—that is, knowledge keepers use their entire bodies to store information.

The Australian Dyirbal language group song specialists sing their kinship system, their social relationships and responsibilities, all about birds and mammals and fish and insects and reptiles, and about techniques for hunting and fishing. They sing the places across their own territory and beyond, the causes of weather conditions and illnesses. They sing about love, grief, bravery and historical events.[16] The Haya people, in north-western Tanzania, sing every phase of iron working in order to get the rhythms and timing of the process right. It's highly memorable because they use sexually explicit rhythms and movements that describe the human reproductive cycle while they're working the iron.[17]

Australian anthropologist John Bradley recounted being taught the songs of the Yanyuwa, in which the many verses described all the land, sea and freshwater species of animals and plants. He described the experience as 'a biology lesson in sung form'.[18]

I had to revisit a decade's research on Indigenous knowledge systems to identify the specific skills being employed. I was astonished at how many there were. Performing knowledge reliably, such that an entire population can learn from it, is not a simple task. I realised that Elders had to be skilled not only in musical and dance skills, but also in cognitive, language and social abilities. The skill set grew.

Knowledge, not wealth, brings power

So who is memorising all this knowledge? Those who control knowledge in oral cultures are almost invariably in the roles of

Elders, and extremely powerful. Hunter-gatherer-fisher communities are often described as egalitarian. In my many years of
researching Indigenous knowledges, I have never come across
a culture that is egalitarian when it comes to information.
The egalitarian description arose from the superficial observation that there are cultures in which there are no individuals
who display personal wealth in excess of other members of the
community. The evidence from myriad ethnographic studies
is unequivocal. Power in traditional oral societies, and those
in the early stages of farming, is in the hands of those who
control knowledge.

Through initiation, the young are admitted into higher
levels of the knowledge system. Critical information must be
retained accurately if it is to serve the needs of survival. Every
culture I have explored employed formal teaching methods,
using initiation levels to instil the songs and their associated
stories and dances. The higher the level of learning, the more
powerful the individual.

In documentaries describing imagined prehistoric cultures, the intellectual domain is almost never represented.
Usually, the 'primitive' people are shown as living a life bereft
of science and immersed in superstition. In documentaries
on contemporary Indigenous cultures, the implication is too
often that knowledge is just handed on casually around the
campfire from parent to child. In the Indigenous cultures
I have explored the world over, this is never the case.

Western ethnographers who made early contact with First
Nations peoples were usually part of an invading culture and
lacked the linguistic skills needed to understand what they
were being told. The priority of the invading colonisers was to

'educate' (or eliminate) the 'primitive' cultures, and to bring them the so-called blessings of civilisation—in particular their 'civilised' religion. Not only did they wipe out cultural beliefs, often violently, but the supposed imperative to teach writing literally overwrote oral tradition as well. In most of the world, the colonisers have succeeded.

Equally unethical is the tendency for 'New Agers' to mis-appropriate Indigenous imagery and supposed knowledge as they romanticise their personal link to native spiritual mediums. There are hundreds of Indigenous American cultures and there is no such thing as a single 'Indigenous American spirituality', let alone spirit guides. Indigenous people are working with the same brain as everyone else, created by the same genetic coding. Scientific knowledge is known through study, obser-vation and learning. These are intellectual achievements.

I have explored the teaching methods of Indigenous Ameri-can, African, Australian Aboriginal, Torres Strait Islander and New Guinea cultures, New Zealand Māori and the Melanesian seafarer schools. All cultures had extensive formal levels of schooling, taught by the knowledge holders. Most involved decades of learning, and a vast proportion of that was rational information, encoded in mythology and performed.

Part of the problem is that we keep hearing Indigenous stories that sound rather simplistic. The reason these sound childlike is that they are the stories told to children to get them used to the characters who populate the more sophisticated stories, which will be taught to them when they grow into adults and reach higher levels of initiation. Without becoming accepted members of the community, we have no right to access anything beyond the children's stories. Unfortunately, the tendency for these stories to be presented without this explanation leads many people to deem the intellect of the adults of the cultures as limited and unscientific. The truth is that those cultures simply wouldn't have survived if this were the case.

Restricting knowledge ensures the longevity of Indigenous stories

People from Indigenous cultures have much information that they will not share: it is secret. Asking for such information is incredibly culturally insensitive and puts the knowledge keepers in a very difficult position. Some knowledge has been kept secret for very good reasons.

Information is very easily corrupted. If you have never played 'Chinese whispers', or the telephone game (as it is known in North America), then I suggest you give it a go. Within minutes, the story that is being passed on will be corrupted. Information will have been lost; surprisingly often, new information will have been added.

What would happen to an Indigenous culture if their survival knowledge could be so easily corrupted? Let's imagine that, generations ago, during a severe drought, their ancestors discovered that a particular fruit they would never normally eat could be made palatable by crushing and soaking it and making it into some kind of biscuit. It doesn't taste great, but it will keep you alive. The biscuit isn't normally eaten because there's usually much better food available. But there's a slight problem. There's another very similar fruit, closely related to the first but it will kill you no matter how you prepare it. This information must be stored really accurately and really reliably.

Cultures without writing ensure that their memorised knowledge is kept accurate by restricting it to a few initiated people, who regularly check with each other that they still remember it correctly. They repeat that information in restricted circumstances in a regular cycle. This is a survival mechanism, and can be seen in cultures the world over.

For example, restricted knowledge accurately retains critical information about how to survive when the predicted migrations of caribou or whales fail to appear for the northwest Alaskan Nunamiut and Tareumiut peoples, respectively.[19] Strategies are far more complicated than just finding another food source. They include complex agreements with trading partners, recognising climatic indicators, storage and kinship responsibilities, and using alternative hunting methods learnt from other tribes. Knowledge learnt far in the past ensures survival far into the future. But it must be kept accurate and therefore is kept highly restricted, and is constantly checked and rechecked. Similar complex survival knowledge is stored at highly restricted levels in Indigenous mythology globally.

When working with the First Nations people in Australia, I was constantly aware of the protocols regarding 'secret men's business' and 'secret women's business'. I soon learnt that I must not ask for more information than I was being offered or that was already in the public domain. It's a tough call on Elders, being asked to give away knowledge while desperately trying to hold on to traditions dating back thousands of years. For them, the revelation of secret knowledge can be severely punished. Distressingly, there is also the concern that if the knowledge isn't recorded and the young people don't take up the tough learning and initiation, then the precious scientific and cultural knowledge will be lost forever. It is painful to watch Elders dealing with this dilemma; it must be hurting them so very much more.

As access to secret knowledge can only be given to those initiated into the higher levels of the culture, it is rarely granted to ethnographers. Men's knowledge is often kept separate from women's knowledge, enabling more to be retained and further levels of security to be implemented. Consequently, early Western male ethnographers—and most of them were male—were not told of women's knowledge; many did not even know or suspect that such information existed.

Given how easy it is to corrupt a very short story in just a few minutes, it is astounding that we have robust evidence of restricted information held by contemporary Indigenous knowledge keepers that dates back well over 10,000 years. Yes, I really did mean over 10,000 years. When you contemplate that Stonehenge is only 5000 years old, it is almost inconceivable that some of the stories date back more than twice as long.

Recent research by geographer Patrick Nunn conclusively demonstrates that Australian Aboriginal cultures retained accurate knowledge of coastal landscape transformations due to sea-level changes and of volcanic eruptions over at least the past 10,000 years. Nunn also argues that the effects of post-glacial land submergence in north-western Europe are stored in stories that have endured for 5000 to 15,000 years.[20]

Cultural astronomer Duane Hamacher has also documented many stories of astronomical events from Aboriginal and Torres Strait Islander Elders. One tells of a meteorite impact in Central Australia that has been dated to 4700 years ago.[21] The oral traditions of the Canadian Heiltsuk Nation talk about surviving the last Ice Age, 14,000 years ago, a fact only recently reaffirmed by archaeologists.[22] There are many more examples, but this will do to demonstrate just how long and how accurately oral tradition can store knowledge of actual events.

How did oral cultures all end up using the same methods?

The memory techniques identified in oral cultures, on which I will be drawing in the rest of this book, seem to be universal. The same techniques of song, rhythm and dance, along with stories told of vivid mythological characters, are found everywhere.

Am I going to argue that this is because all these cultures had some form of contact with each other, despite being separated by massive distances in both space and time? Of course not. There is one obvious common factor. All these techniques depend on the way the human brain works. It is as if these cultures have studied the most recent neuroscience and applied it in designing a knowledge system. The reality is that the cultures which optimised the use of their genetically granted skill potentials are the cultures that survived and flourished. Neuroscientists are only now learning why.

It turns out that not all cells are the microscopic things I had always believed them to be. Some of our human neurons are over a metre long.[23] Neurons are specialised cells in the nervous system that transmit information throughout our bodies.

The structure in our brain that is critical to learning and memory is the hippocampus. It consolidates short-term stored information into long-term memories, which is absolutely essential if we are to develop our abilities beyond simply eating, sleeping and procreating.

What is astonishing to realise is that whenever you think about something new, as you are doing at this moment, new neural pathways are being created. Physically, new neurons regularly come into being. But they don't hang around unless you repeat your consideration of that knowledge. And they don't hang around if that knowledge is boring. New, creative, novel ideas will stick around in physical neural pathways much more reliably than dull stuff. So different stories with lively characters doing wildly exciting things, particularly if they are violent or vulgar, will stay in the memory far better than a list of the properties of the plants you are trying to remember. Turn the plant into a character, though, and its life cycle into an entertaining story that can then be performed, and the knowledge will be securely stored in memory.

The neurons communicate to each other chemically across the gaps between them, the synapses. Your brain strengthens the synapses that are most active and lets those that you don't use fade away. Repetition of knowledge is essential, which is why ceremonial cycles and constant repetition of songs, dances and critical stories are prominent in all Indigenous cultures.

Unlike the constant boring repetition of rote learning, repeating knowledge in a performance is much more enjoyable. Even though the skeleton of information stays the same, every time you dance it or sing it or elaborate on the stories and characters, it is a little different. Performance also requires concentration, making it even more memorable. Your brain is excited by physical movement, music, repeated rhythms and all the aspects of these Indigenous methods I have mentioned so far. Obviously, this brief dip into neuroscience is a gross oversimplification, but it will do for our purposes.

Recent neuroscience has shown that your brain is also particularly good at associating knowledge with place—and that's our topic in the next chapter.

CHAPTER 4

Why oral cultures mark their landscapes

So NF1 is implicated in our spatial abilities. But is this spatial advantage so valuable that it was part of the reason evolution ensured that every one of us has the fragile human allele?

As described earlier, I was astounded by the way I could encode vast amounts of information through testing the way Indigenous cultures turn their landscapes into sophisticated memory spaces.

As I reflected on the physical devices used by Indigenous cultures, I started to recognise how much more was involved than the simple spatial skills I had identified in my early discussions with Andrea and Vic. The list of skills needed for Indigenous knowledge systems just kept growing, and I no longer knew whether NF1 was implicated in enough of them to justify arguing that knowledge systems drove evolution to embed it in every human genome. Before we could make that assessment, I needed to review my research on Indigenous methods that involved visuospatial abilities and think about the skill set involved.

Physical memory devices in Indigenous contexts

Spatial abilities are critical to the way Indigenous cultures engage with art and their landscape. Oral cultures the world over use the landscape as a huge, complex and multilayered memory palace. The same cognitive basis leads to Australian Aboriginal songlines, Indigenous American pilgrimage trails, Pacific islands ceremonial roads, Inca *ceques* and African sung paths. Similarly, portable devices and marks on rock faces, often described as art, serve a much more complex role as mnemonic devices.

I have been profoundly influenced by Elder and storyteller David Kanosh, from the Tlingit tribe of south-eastern Alaska. His insights, along with those from Australian First Nations Elders and others globally, greatly enhanced my ideas about the physical mnemonic technologies in terms of NF1.

'Sometimes anthropologists find me and quiz me on some things,' David told me. 'None had ever asked how I was able to remember so much information on plants, animals and animal behaviour, clan histories and so many other things.'

When I first asked that key question—'How the hell do they remember so much stuff?'—I read all the research I could find on 'primary orality'. But when I started working with Indigenous colleagues, I discovered more. There was so much more.

Indigenous cultures across the world use a vast array of

physical devices, ranging from the landscape, sea and sky to tiny carvings. These devices were not included in the research on primary orality. They gave me a rational, scientific, pragmatic ticket into the fascinating world of archaeology, so dependent on physical remains.

An Indigenous culture may hold thousands of songs, poems, dances and narratives that together form a knowledge system. This repertoire will be constantly updated with new information or refinements on existing knowledge, while losing information that is no longer valued. If the internet were not indexed, you could not search for any information because it would disappear in a quagmire of data. Similarly, oral tradition is indexed, but that index is usually held in the form of physical technologies.[1]

These material mnemonic technologies, implemented according to the resources available, are found across all the oral cultures I explored. They include the landscape, skyscape and seascape, decorated posts, figurines, memory boards, notched staves, decorated utilitarian objects, knotted cord devices, sets of objects and a vast array of artworks. This chapter can only glimpse their range and implementation, but understanding these physical artefacts became essential for me, and I explored the archaeological record to identify the wide span of abilities granted to humans by our ancient knowledge gene.

Understanding the use of landscape for memory

A starting point for understanding the way in which Indigenous cultures use their landscape to store knowledge is via the ancient Greeks. They documented a simpler method, which they called the method of *loci*.

To implement this method, think of a series of physical locations you know well, perhaps around a building or down the street. At each location, you associate information you wish to remember. If I walk around my house and garden and then down to the shops and back home, I will walk past locations I have assigned to every country in the world, ordered from the largest population to the smallest. If I walk in a different direction around the streets near my home, I will cover all of time, from 4500 million years ago to the present. I have a mental hook ready for every place on Earth and every moment in time to add more and more information.

Your brain is really good at associating memories with physical locations. Your hippocampus becomes very active when engaging with spatial knowledge, so deliberately making a connection between a piece of information and a location ensures that the two concepts become linked in your neural pathways. Thinking about the location will trigger the information, and thinking about the information will recall the location (and everything else you have encoded there).

The method of *loci* is associated with the Greek poet Simonides of Ceos, who, according to legend, was able to recall where people were sitting at a banquet before the roof of the hall they were in collapsed. But more sophisticated versions of the method of *loci* existed within oral cultures long before Greek orators mentally walked their streets and buildings as they presented their lengthy speeches. We know this through the stories and memory methods explained by contemporary First Nations societies with continuous cultures dating back well over 10,000 years. I suspect that the Greeks were adapting the system that their pre-literate ancestors had used.

From around 1200 CE, the non-literate Inca briefly conquered lands from Ecuador to Chile, the most extensive empire in the world at that time. It extended further than that of their northern neighbours, the literate Aztecs and Maya. The Inca used a sophisticated version of the method of *loci*. Their capital, Cusco, was surrounded by over 40 imagined pathways, or *ceques*, with over 300 shrines, or *huacas*, along the way. The *huacas* acted as memory locations, as recorded by the Spanish invaders during the 16th century, who then went on to destroy it all. We'll return to the Mesoamerican cultures in Chapter 10.

The Indigneous American Pueblo peoples have a pilgrimage trail connecting one of their language groups, the Zuni, with a location in the Bandelier National Monument. For hundreds of kilometres, the names of the shrines describe the ancient migration routes; they are still recited today. In the Middle East at the beginning of the 20th century, archaeologist Gertrude Bell rode with Arab horsemen who named every feature in what seemed to her to be an almost featureless landscape.

Oral traditions the world over link knowledge to streams, rivers, rocks, lakes, cliffs, hills, trees and other features in their natural memory palaces. The songs, stories and associated knowledge are recalled whenever the landscape is walked, either in reality or in imagination.

Australian Aboriginal songlines structure knowledge

Australian First Nations have a continuous culture dating back at least 65,000 years. The English terms *dreaming* and *songlines* are frequently used to describe the knowledge systems for the hundreds of Aboriginal Nations.

Songlines, or Dreaming tracks, are narrative pathways physically mapped across the entire continent and connect songs and ceremonies to each sacred location. Australian Aboriginals use the term 'Country' to refer to a landscape loaded with cultural knowledge. Over 800 kilometres of songlines, for instance, have been mapped for the Yanyuwa people, who refer to them as *kujika*.[2]

A songline is often a complex mesh of locations at which music, dances and stories are performed to encode a vast encyclopaedia of knowledge. Critically, songlines are not used by individuals alone, but by communities.

Australian First Nations cultures use a sophisticated version of memory palaces. I took advice from my Aboriginal Australian colleagues—especially Margo Neale and Tyson Yunkaporta[3]—and added some of the extra bells and whistles that differentiate the simple method of *loci* from the far more complex sung narrative trails used by Indigenous cultures the world over. It was time to make my Chinese memory palace sing![4]

Tyson insisted that I should try to include people. I couldn't see at first how to do this, but I tried. If someone appeared outside a house while I was staring at their home, adding a new word or character, then I explained what I was doing. People wanted to chat. I'd tell them what I was encoding and they invariably made some personal link.

Anna was in her garden at the location I had chosen for one of the Chinese radicals, the indexing system for the characters. Each radical ends up with huge numbers of words attached to it. I needed to encode the word for 'love'. Anna started telling me that her great love was Australian native orchids.

She became part of the story for her location, as did her orchids—which made it a far richer location in my mind, and so easier for me to encode further information (see Plate 4.1).

When I want to recall something, I just jump to the songline location in my imagination. Eventually, I don't need to do this to recall words and characters I have encoded—they become automatic. But I always use the songline for new information.

The best known of the Aboriginal songlines is the story of the Seven Sisters, which has varying versions across many Australian First Nations cultures. The songline tells of the sisters' battle with an Ancestral Being, Wati Nyiru, who takes the form of a man pursuing the sisters in order to possess them. During their journey across the heart of the continent, he shape-shifts into temptations such as foods, shade or water. The story conveys information about items the women (and therefore all people) need to survive in the desert. The narrative tells of obsession, trickery, passion, loss, survival, rape and evasion. The highly memorable story teaches about laws and expectations, kinship, sustainability, codes of behaviour, the environment and survival, and so much more. Eventually, the sisters throw Wati Nyiru, who has taken the form of a snake, into the sky, where he becomes what many of us know as the belt of Orion. The sisters leap into the sky and can be seen as the star cluster known as the Pleiades, linking the skyscape to Country. All physical space is interconnected.

Given the prominence of the Pleiades and the consistency of this form of knowledge keeping, it is not surprising to find similar First Nations stories the world over. For the Indigenous American Dakota, the stars represent the Seven Girls who were taken from a camp by a red eagle. Much of the Dakota star knowledge depends on a similar framework of mirroring what is in the sky to the land below, known as *kapemni*. A cave in Saint Paul, Minnesota, on the Mississippi River, links to a specific star, as the *Dakota/Lakota Star Map Constellation Guidebook* explains:

> 'We Dakotas come from a cave or star in Orion's belt or we have always been here.' [Jim] Rock believes these three ideas are simultaneously true because the cave by the river mirrors the star by the Milky Way. Because of this sky-earth mirror principle, the *Wakan Tipi* cave by the Big Waterfalls River (*Haha Wakpa/Wakpa Tanka/Misi Zibi*) mirrors the Orion's belt origin star (in the *Tayamni* constellation) by the Milky Way Road of Spirits. For Rock, the tipi poles lean together to connect and point us from specific places where the river joins on Earth Mother below to Sky Father above, so we always live together as good relatives.[5]

First Nations cultures link the sky and land, but also depict their knowledge in a glorious array of art forms. All are integrated, each reinforcing the other. We can see this best through a case study of a single culture. It is time to meet the Tlingit.

Introducing the Tlingit of south-eastern Alaska

The Tlingit are Indigenous to the Pacific Northwest Coast of North America, with a territory covering more than a thousand islands. The oldest dated stone fish weir ever found in the world is in Tlingit territory, estimated to date to at least 11,100 years ago. It is now approximately 52 metres below sea level. The ancestors of the Tlingit must by then already have lived there long enough to understand salmon migrations and to develop weir technologies perfectly suited to harvesting the fish.

The coastline was once way out on what is now the continental shelf. After the end of the last glacial maximum, the sea level was actually above modern levels. Starting around 12,000 years ago, the sea level rose over 53 metres in a 1000-year period. This is an average of over five centimetres per year. This means a massive flooding of the land surface every year, so it is not surprising to find that flood narratives are common in Tlingit culture.[6]

Tlingit is also a contemporary culture, integrated into the modern world. The extraordinary Sealaska Heritage Institute was founded in 1980 to perpetuate and enhance Tlingit and other south-eastern Alaskan cultures, the Haida and Tsimshian. You can get lost for hours exploring its engrossing website.[7]

The Tlingit are divided into two social and ritual groups (known as moieties), Raven (*Yéil*) and Eagle/Wolf (*Ch'aak'/Ghooch*). They identify with their matrilineal clan. Intellectual property laws cover not only stories, names, songs and dances, but also decorated objects ranging from the famous totem poles and clan house posts to blankets, carved horn spoons and wooden storage boxes. Artistic designs, as well as the physical items on which they are found, are owned and can only be used by those with clan permission.

Remarkable conversations I had with David Kanosh gave me insights into this culture that I could never have grasped just by reading books and academic reports. I had mentioned Tlingit totem poles before in my writings, but only briefly and without the nuances that I gained from talking with David. This reinforced yet again that it is only through direct contact with Indigenous leaders that we can hope even to glimpse the complexities of the way knowledge is stored in these cultures.

Throughout the next section, with his permission, I will include David's own words from those conversations on Tlingit knowledges and the way they use physical memory cues for knowledge.

I am of Raven moiety, clan Deisheetaan (Village at the End of the Trail).

As the Tlingit people continued spreading across southeastern Alaska and the Tongass area in British Columbia, one group found a beaver's trail and followed it and at the end of the trail they made a village. This is why we use the emblem of the beaver as our clan crest.

My maternal grandmother and my family were preparing to get on the ferry at Killisnoo. She and Grandpa George looked around the area. Grandma Mary started pointing out places in the surrounding area, telling me the names of some of these landmarks. Throughout the ferry ride to Sitka, Grandma Mary pointed out the names of so many places and commented on the area's outstanding features.

Once home, Grandma Mary started telling me a story with some strange characters and their sometimes over the top antics about each spot along the journey we had just taken on the ferry with my Grandpa George listening on this narrative and smiling.

Grandma Mary did this story narration quite a few times when she visited Sitka.

A research team, consulting with tribal governments, studied the alpine cairns on Baranof Island, the island in south-eastern Alaska containing the city of Sitka. Of the 50 cairns recorded, four were carefully dismantled, examined and reassembled under Indigenous supervision. There were no associated artefacts, so these were not burial places for people or objects. A combination of dating methods showed that the cairns were prehistoric, built within the last two millennia. Anthropologist and oral historian Dr Thomas F. Thornton worked with Tlingit Elders using physical, historical and oral history to conclude that the cairns were built by ancestors of the Tlingit—and, more specifically, by ancestors of the Sitka and Kootznoowoo tribes.[8]

The Tlingit had created memory palaces to aid the knowledge being retained thousands of years after the migration events occurred. David described exactly how he was taught the stories of cairns:

> I will walk to the rock cairn.
>
> I will touch each stone.
>
> I will recite the stories of the medicinal plants . . .
>
> The rock cairns which are atop a few mountains were special places where my Grandfather George John Sr took me to recount stories.
>
> As we went up the trail to one rock cairn, Grandfather George recounted the story of the flood as we passed certain points that were memory spaces for this story. As we arrived on top, I sat down, unable to believe what I was seeing.
>
> 'When the flood water started rising, our people built these to help save us and to preserve our knowledge,'

Grandfather George said quietly. 'The stories I will tell you here are not for everybody.'

My little legs ran from stone to stone as he touched each stone and told part of some incredible stories. Once completed, he quizzed me on what he had spent a considerable amount of time recounting. Each day, we made the rounds three or four times but he paused in the narration and signalled for me to 'fill in the blank' where he left off.

Each day the 'blanks' that I was supposed to fill in got bigger until by the 7th day, I completed the narrations from beginning to end without fail. The narration process was repeated in the similar fashion until I was able to tell the stories correctly from beginning to end . . .

These are ancient stories, stories which go back to the time of legends for these are the stories of the Tlingit . . .

Even as I write this now, I see Raven and his uncle Yook'iskookeik fighting. (That name is so familiar to me because that is my name.) I see the Tlingit going up the mountains, carrying everything they needed to survive. I see the Tlingit putting the rocks together. I see the Tlingit jumping into canoes and look at the waters rise around them higher and higher higher higher. A world which had seemed so stable to the Tlingit now seemed so uncertain. Everything was changing dramatically from what the Tlingit had known.

Art in all its glory just adds more to Country

On the other side of the world, in Pitjantjatjara territory in South Australia, can be found Cave Hill. More correctly known as Walinynga, it is the only known rock art site related to the Seven Sisters songline. Among the multitude of native animals depicted are wonderfully accurate images of horses and camels, which arrived with Europeans in the last few hundred years. Like a record of knowledge in any society, the rock art is regularly updated with new information. Indigenous adaptability is the story of all of human existence.

Australian Aboriginal cultures are often referred to as the oldest continuous cultures on Earth. According to anthropologist Howard Morphy, who works closely with Aboriginal communities in the Northern Territory, the artwork in western Arnhem Land around Gunbalanya (formerly Oenpelli) is the oldest, most continuous and most complete art record anywhere in the world. Over a long timespan, the art changes in a continuous sequence, reflecting the transformation of the landscape.

Dating to before the end of the last Ice Age, around 15,000 years ago, the oldest paintings at Gunbalanya reflect the location's inland environment. Gradually, while retaining constant themes, the images reflect the area becoming coastal and tidal, including the flora and fauna of freshwater floodplains. Eventually, the art changes into the contemporary western Arnhem Land motifs. During the last few thousand years, the distinctive, powerful 'X-ray art' dominated. Although the same animals had appeared before, their defining features now included their internal organs and bone structures; this information had been encoded as knowledge. These 'X-ray artists' would record the arrival of Europeans in the same style, with their horses, buffaloes, boats, guns and strange clothing.

The Gunbalanya sequence dates to long before the Neolithic archaeological periods, demonstrating that knowledge has long been stored in this way through art. It is not, therefore, far-fetched to argue, as I do later in the book, that these knowledge systems can be traced back to the time when humans first moved across the Earth. It is not far-fetched to claim that such knowledge systems are fundamental to our being human. Our success as a species depended on them. And it is not far-fetched to make the link between this aspect of knowledge systems and the human allele of the NF1 gene. But I am getting ahead of myself.

First Nations artworks all over the world include images of animals in realistic and stylised forms. They also represent animals with human features, as many animal ancestors morph into human forms in stories. This enables the story to encode both expectations of human behaviour and details of the animal itself. The landscape, animal tracks, human activities and plants are common motifs.

It is also important to note that a large proportion of Indigenous designs consists of dots, circles, spirals, chevrons, arcs, lines and a huge variety of other abstract forms. Abstract designs allow a multiplicity of meanings to be encoded into the same image. Complexity can be added as knowledge is taught at more and more restricted levels. Information that must be kept accurate is usually restricted, and tends to be encoded with abstract symbols. Abstraction also enables a highly adaptable knowledge system. Meanings can be constantly updated and enhanced in a way that is far more difficult with representational symbols.

Iningai Traditional Owners are working with archaeologists to catalogue the art displayed at Marra Wonga, about 890 kilometres north-west of Brisbane. In the Iningai language, *Marra Wonga* means 'place of many stories'. The rock shelter is 160 metres long, and is estimated to be adorned with over 15,000 petroglyphs. Traditional owners refer to the site as a major teaching site.

Teaching sites are not museums. They are dynamic places where oral tradition is regularly performed, where knowledge is updated and where the physical signs of that knowledge are adapted accordingly. The rock art at Marra Wonga dates to over 5000 years ago but was still being updated only a few decades ago.

It is worth noting, for those who consider agriculture to be a massive leap forward, that first contact reports of the Iningai hunter-gatherer people describe them as reaching over six feet tall and having a life span of 90 years.[9]

Rock art is a key indicator of a knowledge space.

Art is added in massively impressive ways

At archaeological sites across the world, post holes indicate timber posts that have long ago rotted away. Although some posts may have been used for buildings, those discussed in this book were erected to stand alone. Among a number of signs that archaeologists use to tell if they were once part of a building are indications of dripping from the roof. Why would they have erected poles that, the archaeologists assure us, were not used to support roofs? The answer is fairly obvious when we mention that superbly decorated poles also play a significant role in contemporary Indigenous knowledge systems. Decorated posts are found across the world, including in North America, New Guinea and Australia, along with the posts in meeting houses such as the *marae* of Polynesia.

The best known of these, and incredibly spectacular, are the totem poles of the tribes of south-eastern Alaska and British Columbia.[10] Here I return to my fascination with the Tlingit.

Tlingit people sculpt clan crests on totem poles with figures symbolising characters and events from the mythological past as well as historical events relating to known ancestors and living people. These story figures flow down the pole, from top to bottom. When the pole was completed, there would be a potlatch ceremony, a complex and elaborate feast at which guests would hear the stories encoded in the pole. As is the norm for Indigenous cultures, the stories were often performed with costumes, drama and songs.

Carving totem poles is a continuing cultural practice (see Plate 4.2).

There are many stories associated with the Loon totem poles. One relates to when Tlingit and Inuit fishers, among many others, head out to fish or travel. They recall their stories and songs of a fairly nondescript yet highly significant little bird. The loon or diver (*Gavia* spp.) always returns to land at night, calling the whole way back. Should a fisher become disorientated for some reason, such as unexpected bad weather, then as long as they can identify the piercing call of the loon, they can follow it back to land.

But it isn't only the famous totem poles that tell these stories. Tlingit tribes historically built plank houses made from cedar; today they call these clan houses. Usually square or rectangular in shape, the clan houses had front-facing designs and totem poles to display the clan and moiety to which the makers belonged (see Plate 4.3).

David Kanosh explained exactly how he used the totem poles and house screens to recall the stories:

> For the art work, it is interesting to know that my grand-mother Mary John, my grandfather George John Sr and myself would look at a totem pole and use very different spots to tell the same stories. We picked out features which stood out for us personally.
>
> In the traditional clan house, there would also be found a giant backdrop also known as a house screen identify-ing what clan house it is. House screens are generally big enough that I encode the story again, encode historical events but also names of people. House screens I 'read' from right side to left side. I was told to encode in a way that is comfortable for me so this is the format I choose for house screens . . .
>
> Here I am in this video[11] telling a very abbreviated history of the Kaagwaantaan clan (Wolf/Eagle moiety) in English language for the first time. Normally I tell such stories in the Tlingit language so doing this in English for the first time sometimes felt awkward but it is done. The giant backdrop which you may perceive served as my memory place for the histories . . .
>
> There were 28 sections to the Origin story of the Kaagwaantaan (Wolf/Eagle moiety) clan which I could have told but given the time limitation, I picked out only a few to tell. I scanned around the giant backdrop to see which part of the story I would tell next.
>
> Kaagwaantaan is perhaps of the most ancient clan lineage of the Wolf/Eagle moieties which had many groups break

off to form new clans thus the house screen behind me has no genealogy included.

Kaagwaantaan has several clan houses and each of those clan houses has a unique origin story relevant to that house. A traditional clan house may have four or more house posts, depending on the size of it and on those posts would be elaborately carved and painted totems quite often relevant to that clan house and its history.

My mind is doing somersaults trying to find all the English words to use here.

Plate 4.4 shows a still from the video David mentions. He is with the Tlingit group Sitka Kaagwaantaan Dancers, performing at the Sealaska Heritage Celebration 2018, telling the origin story of the Kaagwaantaan. Oratory, dancing, clan regalia and the artwork together form an integrated whole for maintaining and sharing knowledge. Plate 4.5 shows dancers of the Shangukeidí (Thunderbird) clan, also known as Chilkat, with the Lukaax.ádi (Sockeye Salmon) clan, also known as Chilkoot.

During performances of the stories, creating an ephemeral image reinforces memory through the process; the final product is of no consequence. In Australian Aboriginal cultures, it is common for storytellers to draw on the ground. Symbols are drawn with fingers, while leaves, twigs and various objects are used to embellish the canvas. This 'sand talk' adds complexity and commentary to the story being related. At times, these images remain during the telling, creating a dynamic canvas stretching over a hundred metres. This scroll of scenes functions very like the narrative scrolls from Asian cultures, which we'll meet in Chapter 10.

Portable art serves as a mnemonic device

Body painting—sometimes temporary and sometimes in the form of permanent tattooing—is used for ceremony, identity and mnemonics in Indigenous communities across the world. However, permanent portable art forms are also found globally, and these became my obsession for over a year when I first discovered their existence.

I have written extensively about the vast array of decorated objects that serve as memory devices for Indigenous cultures. Indeed, I have yet to find a culture that does not use portable objects to encode knowledge. Most will use different objects to encode the same or similar stories, each reinforcing the other. This ensures a robust knowledge system, even when the culture is mobile or forced to leave its land-based memory paths. Portable mnemonic objects are made from whatever materials are readily available. Their creation and use are heavily dependent on spatial abilities.

Masks are part of the knowledge system of cultures around the world. For example, Pueblo ceremonies involve hundreds of masked mythological characters imparting information through wonderfully entertaining dances. These characters, called *kachinas*, are also represented on pottery and in art. In the form of figurines, they are given to children. These are not dolls but a method to establish the foundation of the knowledge structure that will be constructed around the *kachina* over a lifetime.

It is worth diverting for a moment to consider the creation of such objects in terms of our uniquely human skill set. Other animals use natural materials to make objects. They strip twigs and branches to make tools. They use materials to create shelters. In all cases, the form of the original materials is still visible. Only humans have taken a block of wood or stone and sculpted it to form a figurine. Only humans have taken a blob of clay and used it to represent a human or animal. I now believe that is because only humans have the unique spatial skills granted by our knowledge gene. I'll be returning to this theme in Chapter 8, as I seek out evidence of this skill in human prehistory.

Many cultures use corded devices of some kind: woven, knotted or strung together. Best known are the *quipu* (also spelt *khipu*) of the Inca. A main cord holds many attached cords, which hang vertically when used. The complex knowledge system is encoded using different colours, twists in various directions, secondary cords attached to the primary cords, and an array of knots. The device can be constantly updated. *Quipus* were used to maintain numerical records and narratives. This was documented by the Spanish before they destroyed all the *quipus* they could find. Very few *quipus* survived the onslaught.

Indigenous storytellers from a wide range of cultures move objects on a performance space, a mat that serves as a miniature stage on the ground. The objects might be seeds, shells, leaves or any other objects readily available. Arrangements of stones are used by the Inuit to teach astronomy. Pacific islands navigators manipulate stones and stick charts to teach their extraordinary navigational skills. The Highland Maya 'daykeepers', in Guatemala, use sequences of tossed seeds to navigate their extremely complex calendar. The Indigenous American Blackfoot Confederacy use a bundle, which can contain over 160 objects, to ensure that the correct songs are sung in a prescribed order. West African Mende healing specialists manipulate stones that represent various illnesses as the Elder analyses the patient's symptoms.

The Yoruba of West Africa toss sixteen cowrie shells or pine nuts. Specific readings relate to the number of shells or nuts that land face up. This is an extremely complex and restricted system, one that encodes ritual instructions, divinations, navigation, history, laws, rules and expectations for trade and leadership, knowledge of animals and plants, along with a pharmacopoeia. Then there's the even more complex *Ifá* divination system, where the shells or seeds are tossed twice. For every possible combination of face-up or face-down position of the shells or seeds, there are associated verses that tell a story relating knowledge. Anthropologists have estimated that there are over 1200 verses to be memorised at the entry level.

Textiles or skins, beaded, decorated or woven, are adorned with markings to form a mnemonic device. For example, winter counts are used by tribes of the Great Sioux Nation of North America: the Nakota, Lakota and Dakota. These consist of hides or other fabrics on which a symbol is drawn at first snowfall (hence the name) to represent the year just past and the associated knowledge. Many Indigenous American Eastern Woodlands tribes used belts and strings of white and purple wampum shells as evidence of treaties, for storytelling, as ceremonial gifts and to record historical events.

It is worth remembering that objects with engraved abstract designs are used globally. Such objects have been found at archaeological sites across the world, and have generally been labelled as something like 'enigmatic ritual object', with the conclusion that no more can be discerned. That they have a pragmatic mnemonic purpose should at least be considered, given how commonly such devices are used around the world. I'll offer just a few of my favourite examples of what appear, at first sight, to be enigmatic objects until their purpose is explained by Indigenous knowledge keepers.

Sticks, staves, scrolls, boards and the amazing *lukasa*

It should come as no surprise that notched sticks are used as mnemonic aids for a whole range of purposes, by cultures the world over. It is such an easy object to create. 'Message sticks', encoding invitations to gatherings or to resolve disputes, for example, were used widely in Australia and Africa. Notched staves are particularly good for recording genealogies. Well documented are the staves used by Pacific cultures, where each knob represents a generation. This allows the knowledge keeper to recite or sing the genealogy and associated events, touching each of the knobs in turn. Examples include the Rarotongan genealogy staff and the New Zealand Māori *rakau whakapapa*.

Long stretches of birch bark are used by oral specialists such as the Midewiwin of the Ojibwe people (Ojibwa, Chippewa or Saulteaux) of North America. Their inscribed birch bark scrolls cue them in recalling the origin and migration songs that are the basis of their oral tradition. The scrolls include marks to indicate changes in rhythm and tempo, and divisions between song sequences for the performance. Now, we're not talking about a few dozen songs here, as a Western singer might have in their performance repertoire today. Over 1000 songs have been identified in the ceremonial repertoire and are expected to be recounted perfectly by the Midewiwin. Elaborately carved songboards were used in a similar way by the Indigenous American Ho-Chunk (formerly known as the Winnebago) in their ceremonies.

I had the privilege of examining a number of birch bark scrolls and a songboard used by the Ho-Chunk at the Peabody Museum of Archaeology & Ethnology at Harvard, as shown in Figure 4.1.

In Australia, a vast array of utilitarian objects have been decorated with symbols. These include shields, boomerangs and a carved food dish known as a coolamon. While some of the inscribed objects may be publicly visible, others are only for a knowledge elite to see. Many Australian desert cultures use the highly restricted churinga (also called tjurunga or tjuringa). Made of flat, elongated pieces of wood or stone, the object is covered with symbols, which are pointed to as the stories are told. The markings include dots, multiple lines, concentric circles, U-shapes, zigzags, double grooves and animal tracks.

I confess to having a favourite memory device. The Luba Empire, in what is now the Democratic Republic of the

FIGURE 4.1 A birch bark scroll (top) and Ho-Chunk songboard (bottom), drawn from objects examined at the Peabody Museum of Archaeology & Ethnology. IMAGE: LYNNE KELLY.

Congo, created memory boards known as *nkasa* (the singular form is *lukasa*). They are made of carved wood and sometimes decorated with beads and shells. Vast amounts of information were encoded on these mnemonic objects, with each carving, bead or shell able to be read in multiple ways, depending on the context. The *Luba Bumbudye*, or 'men of memory', acted as the tribal encyclopaedia.

I was sceptical that these memory systems really worked as well as the research suggested. So I replicated one and was astounded as I encoded a field guide to the 412 bird species of my state. I would never have attempted this memory feat before. I have since created other *nkasa* and found them amazingly effective. I have watched young children through to older adults experience the same impact. *Nkasa* are astoundingly practical objects.

One of the most profound moments in my life was holding a real *lukasa* for the first time. I had not anticipated that I would have such an emotional response. To my knowledge, there are no *nkasa* in Australia, so it was during a research trip to the United States, after ten years of fascination with *nkasa*, that I was able to hold the Brooklyn Museum's *lukasa* (shown in Plate 4.6).

Imagine my surprise when David Kanosh described exactly the same device from his Tlingit heritage:

As a child my grandmother Mary John showed me how to make a memory board with a plywood board and pebbles and beads which would look very familiar to you.

I took into consideration the full inventory of stones and semi-precious stones which had been gifted to me by an uncle who was a geologist. I started making my own memory boards with all of these by gluing them to several pieces of plywood.

This happened at a time when we were discouraged from speaking our language or sharing our stories of old and yet my teachers praised me for my geological hobby.

For my teachers, I named every stone, semi-precious stone and crystal. For my friends, I recounted the stories of old.

My grandmother never called it a memory board. She just said, 'This is how you will remember things.'

I did note another Tlingit elder who had a board which had beads, pebbles and bits of clay on it. He held this as he recounted stories.

As Neanderthals and modern humans expanded their territories across the world, they marked their landscape with art and created enigmatic objects. An understanding of the way contemporary oral cultures use art forms as memory devices enables us to interpret these activities through a new lens, as I will in Chapter 8.

It was at the same time as I was researching these mnemonic techniques that, on the other side of the world, Dr Andrea Alveshere was discovering a new and uniquely human allele for the NF1 gene. Neither of us could have imagined then how intimately entwined our two seemingly unrelated research projects would become.

By the end of 2021, Andrea, Vic Riccardi and I were very excited about our hypothesis. But was NF1 really so critical in enabling human knowledge systems, or was that just what we wanted to see? So much pseudoscience invades the world because people grab at what they want to see without testing. We had to follow the scientific method and test our ideas. We had to collaborate with other experts to check that we weren't failing to see any elephants in the room. We needed to hypothesise and retest and clarify and check our academic references, and to keep talking with our First Nations colleagues.

So that is what we did.

CHAPTER 5

The oral knowledges skill set

Very early in our work as a team, Andrea, Vic and I were enthusiastically grabbing at any clues that indicated NF1 might be implicated in our ability to learn and retain knowledge. We were well aware that the brain can play tricks, emphasising the connections that it wants to believe. It was time to apply formal research methodology.

Revisiting all my previous research, I created a set of skills on which Indigenous knowledge systems depend. Expecting the set to run to ten, maybe twenty, I was astonished when I was able to identify 89 specific skills that are necessary for Elders to maintain the complexity and accuracy of their pragmatic and cultural knowledge. I sorted those 89 skills into seven categories: music, performance, spatial, art, language, cognitive and social.

I needed to keep reminding myself of the reasoning as I understood it at that stage: over 500,000 years ago a common ancestor of the chimpanzees and humans (and other species) had an ancient version of the NF1 gene. That ancient gene had a lot of biological effects to do with energy and tumour suppression. The human version became fragile, leading to less vigorous tumour suppression for those who had one corrupted copy. But our interest was in the cognitive effects the human allele also handed on.

That meant we needed to focus on the cognitive traits that were affected when one NF1 gene is no longer functioning fully. That information comes from the large cohort of people with the NF1 disorder, who enabled researchers to learn from their experiences. We owe them a great deal.

The various species evolved with very slightly different versions of NF1. Closest to us, the chimpanzees ended up with their NF1 allele. At least three human species—the Denisovans, Neanderthals and us—ended up with the human NF1 allele. What we share with chimpanzees comes from our mutual ancestor. It is not uniquely human. Where we differ from chimpanzees tells us how our genome has encoded uniquely human traits. Only some of those traits are related to our NF1 allele, and these were the traits we were looking for in this research.

If chimpanzees do not display a behavioural trait, such as musicality, then we know that this must be due to something within our 1.2 per cent of coding DNA that differs from that of chimps. To tell if that trait is influenced by NF1, we need to see if it is impacted when one of the NF1 genes is no longer fully functional. That is how we know that NF1 has a major role in encoding for that trait.

The full list of traits for Indigenous knowledge systems can be found in the Appendix. Others might generate a different list, and I expect some will say that I was remiss not to include some particular skill or should not have given such credence to one that I have included. This is science, but not the sort of pedantic science I have been used to when my focus was physics and mathematics. It can't be. Human behaviour is involved, with all its glorious variation and inexplicable idiosyncrasies. The weight of any particular skill will vary with the specific culture being explored, and so the only skills included are those that I found almost universally. If there was a particular trait that was not seen universally, such as the click sounds in some African languages, then that skill was not included.

I sent my list to my colleagues and waited. Did we have a hypothesis worth pursuing?

Working with primatologists, Andrea compared the skills I had identified with those of chimpanzees. We eliminated any skills that might also be demonstrated by chimps. Andrea and Vic then assessed the impact of the NF1 disorder on each remaining skill. Although almost nothing is dictated by a single gene, this process identified those traits that are majorly influenced by our human knowledge gene, NF1.

The results came in

After eliminating the skills that are either shared with chimpanzees or are not significantly impacted by the NF1 disorder, we were left with 45 skills that we can clearly associate with the human allele of the NF1 gene. It wasn't the percentage of skills that mattered, because some skills are more critical than others. What did matter is if the skills that posed challenges for the NF1 cohort would have any impact on the way information is conveyed in oral cultures.

Fifteen of the skills were abilities shared by chimpanzees and humans to some degree. As described in the Appendix, in three cases the degree of difference was so significant that the skills as described were retained as uniquely human. The other twelve were eliminated, reflecting chimpanzee skill in vocalisations, displays and, most importantly, social behaviours.

For example, in the Art category, the ability to visualise a new form from common materials was eliminated. Chimps might strip leaves from a twig, say, to make a tool. However, they will not carve a piece of wood into the shape of a chimpanzee. That is considered a three-dimensional spatial skill and was retained as a uniquely human one.

In the Language category, we noted that chimpanzees do use varied vocalisations that show some degree of prosody.[1] Although they clearly don't vocalise with the same level of complexity as humans, we nevertheless decided to leave prosody out of the analysis.

Many of the chimpanzee skills fell in the Social category. This was also the category that demonstrated the fewest challenges—and even possible strengths—for those with the NF1 disorder. With the aim of keeping our research ruthlessly rigorous, we removed the Social category entirely. That was a pity, as I really wanted to include the social skills of Indigenous Elders, but science doesn't work that way.

If any skill was assessed as not being significantly impacted by the NF1 disorder, it, too, was eliminated from the analysis. That removed 24 skills.

Six skills were assessed as 'unknown', as no research had been done with those with the NF1 disorder to make an assessment. Unless these skills specifically aligned with skills that were clearly impacted by the NF1 disorder, they, too, were eliminated.

To our surprise, fourteen of the skills affected by the NF1 disorder could be seen to offer potential strengths when considering knowledge systems in an oral culture. We were glimpsing autism-like skills associated with the desire to stick to set rules and with enjoyment of repetition. There were also ADHD-like traits, including the enjoyment of participation and social interactions. Fourteen more skills were eliminated.

Most people with autism or ADHD do not have the NF1 disorder—there are many genes implicated in these diversities—but the results prompted new questions. Why had autism and ADHD not been wiped out by evolution? And what about dyslexia? That is the topic of the next chapter.

The survivors

The surviving traits after this winnowing process were those we can say are definitely not shared with chimpanzees, and on which the impact of the NF1 disorder is highly significant statistically. They tell us what potentials the original mutation hundreds of thousands of years ago granted humans. These are:

Music skills

1 Perform music in songs or chants to store knowledge
2 Use rhythm as a memory aid (mnemonic)
3 Add complexity to music
4 Create embodied musical forms

5 Encode knowledge in dance
6 Compose music

Performance skills

7 Maintain core pedagogy accurately, embellish for entertainment
8 Sequence a performance of multiple elements

Spatial skills

9 Conceptualise vast areas of land
10 Visualise distant spaces
11 Navigate from information in songs
12 Use cardinal directions rather than right and left

Art skills

13 Visualise objects in 3D
14 Create abstract art as mnemonic
15 Use abstract elements to enable adaptability
16 Produce standardised designs
17 Create representational art
18 Organise art in ceremony
19 Adapt established artwork for new knowledge
20 Utilise manual dexterity

Language skills

21 Work with language in complex forms

Cognitive skills

22 Retain an integrated knowledge system
23 Use narrative and mythology as mnemonic

24 Use metaphor to reduce mnemonic effort

25 Layer knowledge in levels of complexity

26 Optimise memory skills

27 Integrate environmental knowledge

28 Distinguish between reality and fantasy

29 Use imagination to create characters and scenarios

30 Make patterns with common objects for memory aids

31 Add new classifications of environmental aspects to existing system

32 Encode knowledge efficiently

33 Decode knowledge from encoded formats

34 Maintain complex genealogies

35 Record observations and act upon them

36 Create structured, coherent knowledge formats

If you think back to the chapters on Indigenous knowledge systems, you will see that the correlation of the items on this list with the skills important in those systems is overwhelming. Also take a moment to think through how many of these inherent and uniquely human skills you use. How many of these biologically encoded strengths do we as a society take advantage of in education and throughout life? We'll revisit that topic in Chapter 11.

No other gene affects this combination of skills. Some other genes may contribute to some aspects of these abilities, but non-literate knowledge systems for the vast majority of human existence have employed all these skills as a suite. There can be no doubt that NF1 survived evolution's ravages for the simple reason that it conferred a massive advantage. And that advantage is the uniquely human ability to store, maintain and transmit knowledge in the most memorable ways possible.

Music adds rhythm to life

All Indigenous cultures perform songs and chants to store knowledge, use rhythm to aid memory and encode knowledge in dance, as described in Chapter 3. Accordingly, it is the music category that offers some of the strongest evidence of NF1 acting as our knowledge gene.

Music skills
- Perform music in songs or chants to store knowledge
- Use rhythm as a memory aid (mnemonic)
- Add complexity to music
- Create embodied musical forms
- Encode knowledge in dance
- Compose music

I could find no other gene that has, to date, been implicated in our human music skills. As not everyone with amusia has the NF1 disorder, there must be other genes that influence our musicality. Given the high prevalence of amusia for those with the NF1 disorder, though, there can be little doubt about the critical role of two fully functioning NF1 genes in our musicality.

One of the most respected researchers in the field of paleoanthropology, Iain Morley explored how four groups of contemporary First Nations peoples in four very different environments on three separate continents engaged with music. He described the songs: 'the greater proportion relates stories, descriptions of events, environments, journeys and subsistence sources, and so constitutes an important repository of knowledge'.[2]

In Chapter 7, I will look at the archaeological and pre-historic evidence for music through the lens of a knowledge gene. I'll explore more there about the way music bonds groups of people and impacts our emotions. The important point here is that music is fundamental to all human memory systems that do not offload that responsibility to text.

What seems incredible to me is that there are languages that are so highly tonal they have developed 'drum poetry'. Verbal messages can be sent using drums alone. By changing the rhythms and tones of the drums, actual words are trans-mitted. Drum poetry is best known from Africa, where it offers highly effective communication through the tropical forest environment. These drums are not accompanying any performance and don't use any pre-arranged code. I find that amazing.[3]

Indigenous musicians sing about love and relationships, but also about so much more. How many modern Western recordings would a songster sell singing about the way corn should be planted or how to bind a broken leg?

I tried to understand the use of music in encoding infor-mation when I created a field guide to the birds of my state. As I mentioned in Chapter 4, I have a memory board based on the African *lukasa*, with beads and shells representing all 412 birds. Clearly, the sounds of the environment, includ-ing bird calls, can be replicated much more easily in song than in writing. But I use music for much more than that.

The bird-*lukasa* acts as a miniature songline, with a bead or shell for every location, each representing a bird family. I started by encoding the families in taxonomic order. Later, I would layer the species in each family, but initially 82 families with long and unfamiliar scientific names was challenge enough.

As I first encoded, touching each location as I went, a rhythm emerged. *Dromaiidae, Anatidae, Megapodiidae, Phasianidae, Podicipedidae, Podicipedidae* . . . I found that repeating every fifth family name caused the rhythm to settle and form what seemed like verses. Singing this verse always calls up impressions of the emu, ducks, the mallee fowl, quails and grebes. And so I went on, encoding all the families. In each verse the rhythm became slightly different. Stories formed in my mind from puns related to the family names. Sometimes they instinctively led to dance-like actions, which made the list even easier to remember. The chant became more musical with time, the dances more evocative. I tried to let the music happen naturally, and, to my surprise, it did.

Music offers the perfect medium for replicating the sounds of nature by incorporating onomatopoeia in the songs. With the reproduction of sounds such as bird calls and the repetition of phrases or verses, compositions evolve to encode information more and more powerfully. The Yoruba sing verses that include the sound of horses bolting, floodwaters roaring, dogs barking, snakes twisting in their death throes, the rattle of seeds in containers, the crushing of bones and the ripping of cloth. These songs also make superb use of crescendo and climax, which are extended by repetition and refrains. We can do a lot with writing, but we can't do that!

Rhythm is so critical to memorable music that often the only musical instruments accompanying oral tradition are some form of percussion: clapping hands or sticks, striking drums, thumping on hollow logs, rattling gourds or stomping. This is probably why First Nations songs are sometimes referred to as chants or song-poetry.

Indigenous knowledges are also *embodied* systems. Dances in oral cultures act out mythological events, with many interwoven aspects: pragmatic, spiritual, social, ethical and entertaining. I watched the dancers at the Indian Pueblo Cultural Center, in Albuquerque, New Mexico, perform their famous Buffalo Dance, one dancer dressed in buffalo skin. In the hands of the women were cobs of the ubiquitous corn. Australian Aboriginal dancers are also famous for their ability to engage with the behaviour of animals. As a writer, I could never convey the nuances of the movements of a kangaroo when it detects the presence of a hunter, yet that response can be danced in moments. These dances are more than just entertainment. In both cases, the dances have a critical role in reinforcing the hunt strategies to be used and the animal behaviour the hunters will need to recognise.[4]

Melodious encyclopaedias of hundreds and hundreds of songs are found in First Nations communities across the world. New songs are composed to encode new information, adding to the encyclopaedia. Songs no longer considered relevant are lost, so the repertoire always represents the latest edition.

Music records information all over the world because it is an innately human trick of the knowledge trade. That NF1 is so heavily indicated in our human musicality is a huge tick in the claim that NF1 is our knowledge gene.

Performance makes it all memorable

Indigenous knowledges are performance-based knowledge systems. Being able to maintain the lessons to be taught accurately, while also making them entertaining, is a key aspect. It is also essential to be able to arrange the performance in a sequence. These are the aspects of a knowledge-based performance that are enhanced by two fully functioning NF1 genes.

> **Performance skills**
> * Maintain core pedagogy accurately, embellish for entertainment
> * Sequence a performance of multiple elements

In Indigenous cultures, knowledge isn't conveyed to youngsters forced to sit still at desks, but through performances in which they are active participants from a young age. Rhythm, rhyme, repetition and redundancy help to make the delivery entertaining and paced, so that the facts do not arrive in a single burst.[5]

Performance acts as a reinforcement at any age, an effect that psychologist David C. Rubin describes as 'staggering'; Rubin quotes a study showing that embodied enactment of knowledge increased recall by 165 per cent.[6]

It is not just contemporary theatre that puts on extravagant spectaculars. Professor Howard Morphy has worked mostly with the Yolŋu people of north-eastern Arnhem Land. He described their major yet relatively infrequent gatherings as complex ritual performances, and likened them to contemporary opera in the scale of the effort and spectacle.[7]

To ensure the maintenance and sharing of vast amounts of knowledge, grand performances rely on skills enhanced by the knowledge gene. Elders need to be able to select the stories, songs and dances suitable for any given audience. Some performances are public, while many are restricted, according to age, gender, initiation status and other factors.

People with the NF1 disorder often struggle with the key skill needed here: executive function. For them, planning and ordering tasks can be a challenge. That tells us that the NF1 gene contributes to our abilities for organising and structuring large and small events.

Spatial skills fill the void with knowledge

The spatial skills give us a swag of evidence for the importance of NF1 in the way we can all optimise our ability to learn and remember. The four we identified are the spatial skills strongly impacted for those with the NF1 disorder.

Spatial skills
- Conceptualise vast areas of land
- Visualise distant spaces
- Navigate from information in songs
- Use cardinal directions rather than right and left

A recent meta-analysis of twin studies showed that spatial ability is largely heritable, in the order of 95 per cent. The study drew on a massive sample size of 41,623 same-sex twin pairs: 18,296 identical twins and 23,327 fraternal twins.[8] The study describes, unsurprisingly, the link between spatial ability

and geographical skills, using maps, planning routes, searching visually, identifying symbols and left/right orientation.

Some of the skills related to spatial ability, such as creating
three-dimensional objects, are discussed below when I talk
about art skills. The two skill sets overlap considerably. But
I wanted to emphasise the spatial skills separately because the
use of the landscape as a vast, complex memory palace is universal among First Nations cultures, and central to their oral
traditions. Thousands of songs, dances and stories are archived
in these landscape libraries. The ability to visualise those vast
landscapes, even when distant, ensures the information is
always available.

Traditional cultures use landscapes as story places, associating their sung narratives with a sequence through the landscape
of hills, cliffs, rocks, trees, streams, rivers and lakes—anything
that is a significant feature. Some of the best-documented
examples include Australian Aboriginal songlines,[9] Indigenous American trails,[10] Inca *ceques*,[11] Polynesian ceremonial
roads such as Ara Metua on Rarotonga,[12] and landscapes
across Africa.[13]

Every landscape can function this way because of our
genetically granted spatial skills. In my memory practice I use
natural features, but also houses and side roads, fountains and
gateways, shops, theatres and the Town Hall. Any landscape
will work. That is why I think the ability to conceptualise vast
areas of land should be considered one of the most important,
not only in the spatial category but also in the entire skill set.

The huge value of knowledge embedded in landscapes is
that we are able to retrieve that information readily, whether
we are physically in the landscape or visiting through memory.

Indigenous academic and author Tyson Yunkaporta, a member of the Apalech Clan in Far North Queensland, describes it in this way:

> In Indigenous culture no information can exist unless it is located. The knower is always located within a map/territory of knowledge and story that is profoundly place-based and corresponds with real landscapes. Each point of interest on a path of travel represents part of a story and a repository for knowledge.[14]

Given the spatial skill issues for those with the NF1 disorder, it is no surprise that Andrea and Vic rated navigation from songs as a challenge. They also indicated the significance of this skill because of the negative impact on executive function when the NF1 gene goes amiss.

Navigation often depends on astronomy, once an essential navigation and timekeeping tool, again requiring the decoding from song to reality.[15] The added difficulty with the sky is that it doesn't stay still. Indigenous cultures follow the movements of stars, planets, the sun and moon, and the dark spaces for keeping a calendar and as a memory space for stories. Andrea and Vic were unable to assess whether the ability to track the moving skyscape is impacted when the NF1 gene goes awry, as there is no research on the subject. Similarly, they were unable to assess whether the common practice of linking the skyscape to the landscape would be impacted. Both seem likely, due to the spatial nature of the skill.

A community-based directionality, rather than the confusing individual approach, is common among Indigenous communities. Australian Gurindji speakers use the cardinal directions to describe where things are; for example, 'The flour is to the west of the sugar on the shelf.'[16] My eye doctor related how he had to adjust his normal practice of asking patients to look left and right when working in remote Aboriginal communities. Each of us understands the terms *left* and *right* from our own individual perspective. In a community, wouldn't it be far better to use something that is the same for everyone? And that's quite simple if you use cardinal directions. My eye doctor needed to ask his Aboriginal patients to look north, south, east and west, something he found difficult.

I was sitting in a restaurant in Darwin once, having dinner with a film crew. I asked the Aboriginal sound technician about directions. He looked around, surprised. 'You mean you whitefellas don't even know which direction is north?' he asked. Even if I could have seen the night sky, it probably wouldn't have helped. The Westerners among us had no idea.

Art is more than just beautiful

What is art? Not being the philosophical type, I shall define it the simplest way I can: 'art' is anything that involves deliberate mark-making in two or three dimensions. The word *art* does not imply that the motifs have an aesthetic purpose, although that may be a very welcome secondary consideration.

Even understanding how integral art is to all aspects of knowing within Indigenous cultures, I was still surprised when I identified 23 distinct art skills used by oral cultures to store and disseminate information.[17] Clearly, they weren't all of equal importance when considering the way knowledge has been perpetuated over the millennia. Only eight were identified as skills directly related to our knowledge gene.

> **Art skills**
> - Visualise objects in 3D
> - Create abstract art as mnemonic
> - Use abstract elements to enable adaptability
> - Produce standardised designs
> - Create representational art
> - Organise art in ceremony
> - Adapt established artwork for new knowledge
> - Utilise manual dexterity

The ability to visualise objects in three dimensions is such an important skill in contemporary oral cultures that it is critical to keep it in mind when examining the archaeological record. It is reasonable to assume that art, in the very distant past, was primarily about knowledge but appreciated aesthetically, too. These objects—whether we call them art or tools or enigmatic objects or figurines—give us so much information about humans in the distant past, humans whom we know had two functioning NF1 genes.

As mentioned earlier, birds and primates, among other species, might take twigs and strip off extraneous leaves to create tools with which to extract insects from hollows or honey from hives. But only humans will take rock or wood and create tools that bear no resemblance to the original material object. From boomerangs to spearheads, humans can visualise beyond the lump of material and act accordingly.

Only humans will take wood and visualise the shape of an animal within it, and then carve a recognisable representation. Birds might select mud for nests, and termites build incredibly impressive mud mounds, but only humans will take mud and create figurines, representations of themselves and the animals they watch. And only humans will take in a three-dimensional observation of the landscape and replicate it in a two-dimensional painting or carving.

It is not only realistic forms that are tagged as 'art'. Humans can use abstract elements and associate complex ideas with them. If the horse looks like a horse, it's hard to think of it as anything except a horse. It may be a mythological horse with all sorts of interesting stories, but it is still a horse. Abstract symbols have far more adaptability.

Anthropologist Howard Morphy, in his detailed study of Australian Aboriginal art, describes the way abstract design elements can encode a multiplicity of meanings, depending on the context in which the art is being considered and the level of knowledge of those viewing it.[18] By using abstract symbols, interpretations can be limited to those who have reached a high enough level of initiation to have rights to that knowledge. Restricting knowledge, so critical for accurate retention over thousands of years, depends on the use of abstract symbols. And only humans do that.

One of the most important human traits is adaptability. To move out of Africa and adapt to almost every biome on Earth, our forebears must have had adaptability in spades. That means they also needed adaptable memory technologies, which abstract symbols provide. Their meaning can be changed or the symbol modified as required. As we'll see in

Chapter 8, despite the great love everyone has for realistic cave art, it is abstract symbols that dominate.

The ability to adapt established art to encode new knowledge would have been absolutely essential for cultures in the past adapting to new habitats or changes in their existing territories.

The manual dexterity involved is amazing. In an extreme example, carvings from the Dorset Paleo-Eskimo archaeological sites date back over a thousand years. The tiny ivory effigies of animals were sometimes as small as the head of a match, carved with such precision that similar species can be differentiated accurately when viewed under a microscope. The loon, such an important bird for the Tlingit today, as mentioned in Chapter 4, was clearly just as important back then. Researchers have been able to distinguish the common loon from the red-throated loon even in these micro-sculptures.[19]

Language is the basis of human communication

Although the knowledge gene is clearly linked to the skills mentioned so far, it is hard to assess how much the knowledge gene contributes to language, given we know that other genes, in particular FOXP2, are so important. Prosody was eliminated as a skill aided by NF1 because the vocalisations of chimpanzees exhibit some of the same expressive traits. However, complexity was also considered a potential challenge for those with the NF1 disorder, and not closely related to the language issues associated with FOXP2.

Language skills
- Work with language in complex forms

Indigenous cultures use language in complex forms. Different language is used depending on the access individuals might have to particular knowledge, with formal ceremonial language often differing greatly from that used in everyday life. Memorability is increased by the use of rhyme, repetition, rhythm and alliteration within a song or story. Economy of talk is a huge advantage when all the information needs to be repeated from memory. Consequently, vast amounts of information can be encoded in very specific phrases, words and combinations of words, including an extensive use of metaphor.

These linguistic complexities are sometimes referred to as 'oral technologies' or 'orality'.[20] It was the writings of sociologist Carl Couch that first led me to explore the relationship between primary orality and archaeology.[21] Couch defined 'primary orality' as the oral technologies used by cultures that had no contact with writing. He described these skills as an information technology. I had studied Information Technology at postgraduate level and taught the subject for decades. Using an information technology lens to explore non-literate knowledge systems instantly felt right. Over a decade later, research shows that my gut feelings were justified. As a technology, primary orality increases the amount and complexity of the information a community can store and retrieve.

Cognitive skills are at the heart of knowing

I could instantly see the impact of musical, spatial and artistic skills on oral technologies. A major indicator for Vic was the impact on specific cognitive skills that humans exhibit but chimpanzees don't. The analysis of general cognitive skills led to a rather long list.

Critically, there is no impact from a damaged NF1 gene on general comprehension. This is very important to emphasise, because if there were an impact on a person's ability to understand, it might impact every other skill listed. The whole research project would never have happened.

Although comprehension is not impacted, Andrea and Vic ended up with a suite of fifteen cognitive skills that made a significant contribution to our uniquely human ability to use knowledge.

Cognitive skills
- Retain an integrated knowledge system
- Use narrative and mythology as mnemonic
- Use metaphor to reduce mnemonic effort
- Layer knowledge in levels of complexity
- Optimise memory skills
- Integrate environmental knowledge
- Distinguish between reality and fantasy
- Use imagination to create characters and scenarios
- Make patterns with common objects for memory aids
- Add new classifications of environmental aspects to existing system
- Encode knowledge efficiently
- Decode knowledge from encoded formats
- Maintain complex genealogies
- Record observations and act upon them
- Create structured, coherent knowledge formats

In Western universities, faculties reside in silos. These might overlap a little, but in gaining depth we have lost the big picture that allows us to see patterns and create integrated knowledge systems. The knowledge that can be retrieved from a songline, for example, might simultaneously draw on zoology, botany, agriculture, history, geology, meteorology, climate science, geography, astronomy, natural resources, ecology, theology, music, law and philosophy, among many other disciplines.

Our brains simply love narrative. We are a storytelling species—and those stories serve so many wonderful roles, including storing knowledge and providing lessons related to expectations and values. All forms of narrative, and especially mythology, are used as mnemonic technologies. This purpose requires interpretation from the narrative to decode the relevant information. The stories are often quite complex, encoding interrelated themes. I like the way Fentress and Wickham express it: '[A] story is a sort of natural container for memory; a way of sequencing a set of images, through logical and semantic connections, into a shape which is, in itself, easy to retain in memory. A story is thus a large-scale aide-mémoire.'[22]

I have mentioned metaphors before as part of the use of complex language, but I also included them as a separate skill because they are used extensively by oral cultures as a way of reducing a story to a single phrase.

In contemporary Western cultures, we also use what are known as conceptual metaphors, which reduce the explanation workload: 'Rebecca *kept me in the dark*. She is my *golden girl* but her life was *at the crossroads* as she considered her *life's journey*. Will she *meet her Waterloo*?' These metaphors will only make sense for those who share my cultural background. Outside Western culture, they would be meaningless. When I use these expressions, I am assuming that those I am speaking to understand them, and I won't have to explain further. Oral cultures use metaphor constantly, which often leads those outside the culture to misinterpret the reference, or to dismiss it entirely.

How would you react to being called a tree pangolin by a member of the West African Kpelle language group? Unless you are Kpelle, you would have no idea whether to smile appreciatively or to bow your head in shame. For the Kpelle, a 'tree pangolin' refers to a person who wantonly reveals secrets. The pangolin (a type of scaly anteater) is covered with tough scales except for its soft, white belly. When threatened, it rolls up into a ball, protecting its vulnerable belly with its hard scales. One Kpelle story talks about the way the foolish tree pangolin reveals the secret of its soft belly to a hungry leopard. Not surprisingly, the leopard proceeds to unroll the ball of scales and eat the pangolin.[23]

Vic emphasised the struggles those with the NF1 disorder have with their attention span. The ability to focus, then concentrate and persevere is essential for mastering the memory techniques. I found long periods of intense focus very difficult when first adapting the memory methods I was learning. Everything seemed to distract me. With practice, I learnt how to focus. Gradually, I found the range of memory methods from oral cultures quicker to implement and easier to use.

The strength of methods such as sung narrative paths and portable memory devices is that you can layer further and further levels of complexity on the first level of knowledge.

Once you have the initial sequence physically associated with places across the land, in the skyscape, in art or on a handheld object, you have a grounded structure. You can then add more and more information to each location. Optimising this combination of memory skills requires cognitive gymnastics, practice and, if available, ongoing training from the knowledge experts.

Particularly important within an overall integrated knowledge system is environmental knowledge. In oral cultures, it is usual to use plant and animal behaviour, astronomy and cloud formations in combination as indicators for weather. Geology indicates possible plant species, while plants indicate particular animal species. Those interrelationships depend on intense observation linked with formal learning.

The French anthropologist and ethnologist Claude Lévi-Strauss wrote about Indigenous cultures all over the world classifying hundreds or, in some cases, thousands of species. Ethnographers without a background in science failed to recognise the scientific content of what they were being told. Lévi-Strauss described an ethnographer in Africa who had no interest in botany and so didn't have the vocabulary to understand what the Indigenous people were telling her. In the language used in the 1960s, Lévi-Strauss wrote that the 'natives on the other hand took such an interest for granted'.[24]

This lack of interdisciplinary understanding persists. My husband, Damian, was on an archaeological dig at Lake Mungo, in south-western New South Wales. Although the dig team were united when engrossed in the ancient hearths uncovered in the spectacular windswept dunes, Damian was unable to interest them in the wedge-tailed eagles flying overhead.

His Aboriginal colleague acknowledged them immediately. All present were united in their amusement, however, when some tourists, having travelled the long and dusty track, arrived with a boat strapped to the top of their car. They were 10,000 years too late. The once wetland was now dry. The landscape had changed significantly over the tens of thousands of years the ancestors of the Barkandji/Paakantyi, Mutthi Mutthi and Ngiyampaa people had been living there and, critically, adapted.

I believe the undervaluing of intermeshed knowledge domains is to our huge detriment. We are genetically encoded to make these interdisciplinary links, even though our education system seems designed to keep them separate.

Adult attempts at anthropomorphism often draw derision. Only children, it seems, are allowed to think of animals and plants as displaying human emotions, behaviours and values. Why? It is not as if humans are incapable of understanding what relates to reality and what is imagined in worlds beyond mundane experience. Mythology rules Indigenous knowledge systems. According to Andrea and Vic, aspects of using mythology to encode information are a challenge for those with the NF1 disorder because their thinking has a tendency to be extremely literal.

The constant presence of characters in Indigenous memory systems was emphasised to me by so many First Nations colleagues. In my own experiments, the more I engaged with characters, and the more real those characters became to me, the stronger the stories and the encoded knowledge became. At first, I was concerned by the number of characters I was creating and trying to keep them all in memory, but I soon ceased to worry. When I needed a particular character, they would always just be there.

At first I used people, mostly historical figures. But the more I adapted my systems to match what I was learning from Indigenous cultures, the more I found that animal characters worked even better. They offered so much more in terms of visual differentiation, a wider field for imagery and behaviour and, for reasons I can't fully explain, a different emotional experience. I found that my animal characters naturally morphed between their human and animal forms. I even started including plants, and they gained character just as well. I cannot emphasise enough how much mythological characters enliven, enrich, enhance and add efficiency to knowledge retention.

I am not in the least surprised that mythological beings are represented on every conceivable surface, from rock faces to pottery, in textiles and performances. It would be impossible to overstate their importance and the complexity of their relationship with everyday reality. It is also not surprising that some of the earliest artworks we have are figurines. But more of that in Chapter 9.

Common objects are used extensively as mnemonics for the oral tradition; some are adorned with the images of the characters, and many are simply encoded with the ideas. In southern Nigeria, for example, each culture would use sticks, cords, strings, beads, amulets, tattoos, pots, trophies, marks on walls, erected structures and trees as memory aids.[25]

Arrangements of shells or seeds are commonly employed for stories. The more I read of the Yoruba of West Africa, the more amazed I was and the more I tried to replicate their practice. A knowledge keeper will toss sixteen cowrie shells and immediately recognise the pattern, depending on how many of them land with the mouth up. William Bascom's large book on the topic includes 305 pages of song-poetry, representing the verses known to a single sixteen-cowrie diviner.[26] Although referred to as divination, the verses encoded knowledge of animals and plants, including a pharmacopoeia. They tell of how to protect against smallpox, how to navigate, rules for trading and guidelines for using power in resolving disputes. They include cultural history, legal precedents and ritual instructions.

I bought sixteen cowries and tried it. It did work, sort of, and very slowly, despite the effort I was investing, but the challenge was so complex that I gave up and just sat back in awe of the Yoruba. That awe grew to inconceivable head-shaking when I read Elizabeth McClelland's book describing the even more complex system memorised by the Babalaéwo of the Yoruba *Ifà* cult.[27] They toss sixteen palm nuts twice, cueing the Odù Corpus, which contains 256 sets of ordered verses, estimated to contain over 1200 verses at its most minimal level.

Let's not leave out the Highland Maya here. The 'day-keepers' of the Quiché language group of Guatemala trained Barbara Tedlock to use sequences of tossed seeds as a memory aid for divinations and their extremely complex 260-day and 365-day calendars.[28] I tried this, too. Being a mathematics geek, I loved figuring out the calendars, but beyond that it was again too hard. I needed the sort of training that Tedlock enjoyed. And a lot of time. But I can see that these amazing feats are possible if we use the skills enhanced by our NF1 gene to their full potential.

These mnemonic devices work because the human brain loves patterns and will see them in seemingly abstract objects. Unfortunately, this skill is often a challenge for those with NF1 disorder.

Andrea presented our results on behalf of the three of us at the 2022 conference of the American Association of Biological Anthropologists.[29]

The adaptable human

Our ancestors adapted to new environments and to new knowledge in the habitats they already lived in. The ability to adapt requires the skill of adding new environmental classifications to existing systems effectively and efficiently. It's all very well to have knowledge somewhere deep in the internet or on the bottom shelf of an academic library. The only way we can act on this information is if we can have it available to us in memory, so that we can see patterns and create solutions.

That precious information is not much use unless we can decode it from the encoded formats, which can be an extremely elaborate process. As I've mentioned previously, Indigenous cultures retain genealogies that are way more complex than most Westerners' neat family trees. But why would such complex data sets be retained when they require so much effort?

In 2009, early in my delving into this topic, I consulted the linguist František Kratochvíl, who works with the Abui language group from the central part of the Alor Island, in eastern Indonesia. On my behalf, he asked an Elder and was told that genealogies are used in property claims, where stories relating the names of ancestral owners need to be told within a group for verification. Genealogies are also used to resolve conflicts by identifying relationships. The third purpose given by the Abui Elder was to prevent inbreeding, which they knew would cause birth defects.[30]

Long-term patterns, such as the link to birth defects, can only be detected through constant observation. Astronomy and landscape changes may require detailed observation over years, decades or even centuries. Yet such details are well-known from oral cultures all over the world.

It is not possible to note changes unless you are familiar with what is already known. That is a massive task when you are solely dependent on memory to do so. Was that star in that position at this time last year? Where was the shoreline last year? Did this particular weather phenomenon happen at exactly the same time, and was the wind direction the same?

Only with knowledge comes the ability to adapt, the unique genius of the human species. As this chapter has shown, it is only because of the unique suite of skills granted to us by two fully functioning NF1 genes that we have this complex interaction of skills that, together, give us an ability to store information not available to any other species. The only other disorder that has been suggested to me as displaying a possibly comparable set of impacts is Williams syndrome. Less common (one in 7300 births) than NF1, it is caused by the deletion of 26–28 genes on chromosome 7. This syndrome severely impacts spatial skills and has a higher than normal amusia rating, although only 11 per cent compared with 67 per cent for NF1. But the overall impact of Williams syndrome seems to be linked with a higher rate of musicality and of emotional response to music than the general population, including individuals who are musically gifted.[31]

Much more work needs to be done to show a causal relationship between NF1 and the cognitive responses observed. I will have to leave that to the experts in the appropriate fields.

Although other genes will certainly contribute to the skills we investigated, it is this research into NF1 that demonstrates the critical role of the skills in a combination that is uniquely human and incredibly powerful. For our human creativity and ability to adapt, it is this combination of skills that justifies NF1 being nicknamed 'the knowledge gene'.

There is no doubt that our genome drives the unique set of human skills that grants us the ability to learn, store and constantly update knowledge. The development and analysis of the skill set became the foundation of a new direction in my research into knowledge systems, a direction I would not have taken otherwise.

I'll finish this chapter by again quoting Indigenous academic and author Tyson Yunkaporta:

In Indigenous culture, narrative pathways through landscapes of knowledge are encoded in stories for the production, transmission and storage of information. This is a profoundly human practice, one which arguably makes us human. Knowledge is more readily transferred into long-term memory in the form of a story. All the experiences of your life are processed through story-making.[32]

CHAPTER 6

Questions about neurodiversity

In the previous chapter, I described the analysis of the skill set of 89 items used by knowledge keepers in non-literate cultures. The clear outcome was that a fully functioning pair of NF1 genes is highly implicated in the knowledge systems I had been researching for over a decade. I was delighted. The mutation giving us our unique NF1 gene way back in the mists of time was as valuable as we had hoped. It was evident that the original NF1 mutation was so incredibly valuable that 100 per cent of us ended up with two copies of it, and every chimp-like version had been eliminated from human populations.

In assessing the skill set for Indigenous knowledges against the skills impacted by the NF1 disorder, Andrea and Vic had decided that fourteen of the skills listed were possibly enhanced by the disorder, but only if you considered them in the light of the way non-literate cultures would have stored information in the distant past.

These fourteen skills were very similar to some traits within the autism spectrum and others in the ADHD population. Obviously, not everyone with autism or ADHD also suffers from the NF1 disorder. That just shows that different genes can have superficially similar impacts. The research on autism and ADHD suggests that many genes are implicated in the two disorders, with the full genetic picture far from understood yet. What is clear, though, is that most of those on the various spectrums are carrying a fully functioning pair of NF1 genes.

As inevitably happens when doing research, irresistible questions insert themselves and will not go away until they have been addressed. I now had to think through the implications of identifying strengths when we expected to find only challenges. Evolution has shoved the human NF1 gene into every single one of us. We all have the potential for the knowledge skills that are so highly valued, as long as one of that pair of genes has not been sabotaged.

But what about disorders that only affect a part of the population? Why have autism and ADHD remained in the population, but at much lower rates? The research shows that they are genetically driven traits. If, unlike NF1, it wasn't valuable for all of us to inherit these potentials, why would evolution keep them in just a small, yet consistent, proportion of every human population?

I put aside my NF1 thinking for a while and started asking why evolution had left the genetic blueprint for other so-called disorders hanging around. There had to be advantages. What would happen if I did the same sort of analysis and looked at autism and ADHD as they appeared in non-literate cultures? What about dyslexia, which is so closely tied to writing? And aphantasia (the inability to visualise mental images) and deafness and . . .?

I headed for the academic research. Fortunately, I was not the first to ask these questions, so there was plenty of help available. Talking to many people on neurodiverse spectrums, along with reading the research, led me to conclude that evolution had maintained a sub-population of people with neurodiversities for the benefit of the whole population. Let's look at the thinking behind that conclusion.

Back to evolution and its cost-benefit analyses

As a result of research in this area in contemporary populations, we no longer speak of 'disorder' but 'diversity'. Researchers have explored cognitive strengths for autism, dyslexia, ADHD, aphantasia, congenital deafness and other diversities, and suggest that these should be considered for their value in human evolution. I read their discoveries in the light of our Indigenous knowledge systems skill set, and I talked to many people and their families who were impacted by a diagnosis of these diversities.

As we saw in Chapter 2, the model of sickle-cell anaemia demonstrates that evolution functions as if conducting a cost-benefit analysis for the population as a whole. A knowledge system lens shows that neurodiversities follow a very similar model. This chapter will ask whether neurodiversity, as well as the knowledge gene, was essential for the creativity in thinking that enabled humans to adapt to the huge range of environments across the world in a way that no other species has managed to do.

To solve the world's current problems, should we also be seeking out the diverse thinking that has served us so well throughout human existence?

Thinking back to our neurodiverse past

Our ancestors with the first human NF1 allele genetically diverted from the rest of the population. Hundreds of thousands of years ago, when that first mutation happened, a human became different. The cognitive diversity granted by the mutation eventually became so successful that every individual in every human species for which we have DNA now has what was then a 'diversity'.

As some members of the species gained two copies of the new human allele, they may have been considered by the rest of their social group as having a disorder, due to their distinctive behaviour. It was certainly a divergence—a neurodivergence. Slowly, over what might have been hundreds of thousands of years, the diverse people with a double dose of the fully functioning human NF1 allele were successful enough that eventually all humans inherited their skill set.

But what about other mutations, those that led to what used to be considered disorders? If the genetic diversity appears to be a disadvantage—or, as biologists say, 'maladaptive'—then natural selection suggests it should decrease over time, possibly to be completely eliminated.

So why have autism, ADHD, dyslexia and many other diversities remained in the gene pool? They seem to have a genetic component, and persist despite the difficulties they create for the individuals and their families. It's time to look at these diversities in the context of evolution and non-literate knowledge systems.

It is common for those with 'high-functioning' autism to have overwhelmingly strong specialist interests and the ability for extreme attention to detail. Those with ADHD have an energy level, both physically and intellectually, that can lead them to explore ideas with a diffused, big-picture thinking style. Dyslexics can visualise clearly in three dimensions, rotating and exploring structures in their minds, and are often superb navigators.

It would be easy to wax lyrical about the skills people with these neurodiversities likely brought to a prehistoric population, and neglect to acknowledge that, in contemporary society, these same diversities may cause a great deal of distress and difficulty, especially for those who have extreme forms. I want to emphasise that, although I am focusing here on the strengths these neurodiversities can confer, I am not dismissing the pain that so many neurodiverse people and their families experience.

Were burials for the disabled given special treatment in the distant past?

A number of archaeologists have reported on the occurrence of deformed skeletons in the Late Pleistocene, which is the geological epoch that dates from 129,000 to 11,700 years ago. Some of these articles refer instead to the Upper Palaeolithic, which is the overlapping cultural period that dates from around 45,000 to 12,000 years ago.

According to archaeologist Simona Petru: 'Since finds of deformed and damaged skeletons in Upper Palaeolithic graves are relatively frequent, it may be possible to conclude that people at that time were particularly cautious with the bodies of those who, for a variety of reasons, deviated from normal in their lifetime.'[1]

Anthropologist Erik Trinkaus has presented compelling statistics for what he termed 'an abundance of developmental anomalies and abnormalities in Pleistocene people', concluding: 'No single factor sufficiently accounts for the elevated level of these developmental variants or the low probability of finding them in the available paleontological record.'[2]

At Sunghir, about 200 kilometres east of Moscow, archaeologists have been excavating a hunter-gatherer site dating back over 30,000 years. They found an extremely elaborate triple burial belonging to an adult male, who was covered in beads and ochre, and two juveniles. The male was buried head to head with a ten-year-old and a twelve-year-old, both of whom seemed to have suffered from physical abnormalities that would have been apparent to all.[3]

Were these burials unusually elaborate because they were for people whom their communities valued for their diversity, and who weren't judged purely on what we see as 'disabilities'?

When asking about neurodiversities, I had to return to the type of thinking I had applied to sickle-cell anaemia. In Chapter 2 we saw that evolution was playing a cost-benefit game. It seems cruel, but evolution can afford to sacrifice some people to sickle-cell anaemia because the population as a whole is better off, having many members who are healthy and protected from malaria.

Using the same logic, I looked at other diversities with genetic components and asked: why has the potentially devastating diversity been left in a proportion of the population?

Other researchers have been exploring similar ideas, presenting strengths for autism, dyslexia, ADHD and other diversities that, they suggest, should be considered for their value to human evolution. First coined by the autism movement, the recent term 'neurodiversity' has now been more broadly adopted to refer to those who do not fit the 'neurotypical' stereotype.[4] I will build on this research by asking whether certain neurodiversities were essential to the enhancement of the knowledge systems already being used by modern humans taking advantage of their NF1-granted gifts.

I'm going to present a thought experiment, imagining a small foraging band in our distant past. As ancient foraging groups moved into new territories, they needed to mark their environment, retain their existing knowledge base, adapt that knowledge for new information, perform that knowledge and use it to meet all their practical needs.

Most of our small band will be neurotypical, socially adept, willing to conform, engaging with their NF1 and FOXP2 skills, and valuing their knowledge keepers. A few will be gifted with other invaluable skills that set them apart from the crowd. These gifted individuals include the autistic, ADHD and dyslexic members of the group.

ASD can provide specialised knowledge

According to the Australian *National Guideline for the Assessment and Diagnosis of Autism Spectrum Disorders*, 'Autism spectrum disorder (ASD) is a neurodevelopmental condition. Individuals on the autism spectrum may show difficulties with social and communication skills, and display a variety of behaviours and intense or focused interests.'[5]

Current thinking suggests that autism may result from a range of factors, including genetics, but may possibly also be triggered by biological and environmental factors. There appears to be a very large number of genes that could be involved. Some date back to our shared ape heritage, while others are uniquely human (although still date back over 100,000 years). Some cases of autism may be due to spontaneous DNA coding errors.[6]

The first indications of autism often occur around the age of two or three. The frequently noted signs of autism revolve around difficulty with interpersonal contact. People with autism struggle to ascertain what other people are thinking. They have difficulty reading the expressions of others and perceiving their emotional state. Autistic people often avoid eye contact.[7]

There used to be a stereotype of autistic people: socially isolated, intellectually disabled individuals who do not speak and display repetitive behaviours such as hand flapping and rocking or obsessively lining up toys or watching a moving object. There are also those described as 'high-functioning' autistics; for a while they were identified as having 'Asperger's syndrome', although this is no longer a separate diagnosis. What is of particular interest for the story here is that these high-functioning autistic people tend to have unusually strong interests and an ability to note details the rest of us ignore. They also often display repetitive behaviours, exactly as needed to retain knowledge in the oral tradition accurately and over long periods of time.

I had heard of Tavish Bloom's extraordinary knowledge of Victorian birds from highly experienced local birders, all of whom had built up their skills over decades. If they face a difficult identification, the most knowledgeable convene to debate it, and their number includes Tavish. When it comes to identifying bird calls, the others defer to him. Tavish has been birding seriously for eighteen months. He is twelve years old.

Tavish has been diagnosed as having level two autism, defined in Australia as 'people requiring substantial support'. Our concern here is not with the problems Tavish and his family experience, nor with the support they need. What Tavish taught me is the autistic strength resulting from an intense specific interest.

Tavish explained his obsession with birds to me:

Birds are my main interest. After that animals, and after that nature and climbing trees.

It's important to climb very high in trees. I find it very calming. When people see me so high up that don't know me they freak out. If I'm stressed about school or something, then climbing trees makes me feel better especially if I can watch birds in the tree. It's very cool.

I feel like I've got a connection to the birds. I like being around birds more than people but it depends on the people. I like being around people who are into birds. I'm not that good at social interaction. I spend a lot of time looking at birds and researching them.

I first got really into birds at Deans Marsh [in coastal Victoria]. I saw two Pacific Gulls. They were juveniles and still brown. I thought they were skuas because I didn't know much about birds. I got home and looked them up. I knew skuas from watching tonnes and tonnes of David Attenborough documentaries from three or four years old. I'm fine with real documentaries, even the animals being killed don't bother me. Everything is interesting. I've always been obsessed with animals and birds but then it became birds especially. I built up a library. I've pretty much got all the Australian bird books.

I start thinking about birds all the time. I can see images of them all the time. When I'm doing something like spelling, my brain just starts thinking about birds and drifts off. Pictures of birds just appear in my mind. I don't know why—they just appear. That happens pretty much every time I think about birds, but also when I'm doing other things.

When I was about ten, I couldn't read when everyone else could. When I got into birds, I started reading bird books, so I learnt how to read. My reading is now okay. I can't do long complicated words if they're not to do with birds but I can if they are. The first fiction book that I read from beginning to end was the first in the Guardians of Ga'Hoole series by Kathryn Lasky, which was all about owls.

I learnt most of the bird calls from the Bird ID app but before that from trying to track down the bird to find it. Everyone at school knows that I'm into birds. Some others are also mildly into birds.

I don't mind being not normal. My friend at school has a very strong interest in technology. He's always yabbering away in my ear about movie making. We're trying to put my interest in birds and his interest in technology together to make documentaries about birds. He's getting interested in birdwatching and I'm getting interested in making movies.

It is very easy to see how an intense fascination with animals would be invaluable to our small band of hunter-gatherers.

Other researchers have also considered ASD skills in the evolutionary context. For example, Happé and Frith argue that it is easy to explain why traits featuring an extreme focus on detail have persisted within the gene pool.[8] They see the advantages of being able to identify vast numbers of plants and animals, not only through their physical characteristics but also through movement and sound, and the value of having a unique connection with animals, not least for hunting.

ADHD can grant broad thinking and endless performance

There is no blood test or brainscan that can diagnose ADHD. That is done by medical specialists through analysis of the individual's behaviour. The Australasian ADHD Professionals Association's website says:

> People with ADHD can struggle to focus and concentrate, control their impulses and make decisions that take into account longer term consequences. They can experience difficulties with planning and prioritising, getting organised, and time management. These difficulties can impact the ability to study, work, manage responsibilities, develop and maintain social relationships, enjoy leisure time and relax. They can also negatively impact self-confidence and self-esteem.[9]

The strengths associated with ADHD by author Thomas Armstrong and others include speed when thinking, a need for novelty and change, and relentless physical activity.[10] Contrary to intuitive thinking, people with ADHD are understimulated due to lower levels of dopamine in their brains. They are desperate for external stimulation, which is thought to contribute to their constant movement, impulsiveness and search for new experiences. Consequently, psychostimulants are often prescribed because they have a calming effect.

Armstrong describes how, through family studies, it has been shown that ADHD is a highly heritable diversity. Some studies point to an allele referred to as DRD4, which impacts dopamine production and has been tagged the 'novelty-seeking gene'. The allele is associated with risky behaviours, not just ADHD. It seems likely that DRD4 arose comparatively recently in human evolution, less than 40,000 years ago.

Xander Dark is ten years old and has been diagnosed with ADHD and tic disorder. When talking with other neurodivergents, I was able to say, 'Could you pause a moment while I write that down?' It soon became apparent that with Xander this was not an option. He tried hard to oblige, but in the end we resorted to videoing his answers, which were punctuated by the constant distraction of his cat, Chipsy.

Xander explained a game he has invented to me:

Basically you are in a boat and you want to go round and your first goal is to mine a tree because you start with an axe and then your second goal is to use the money you get from mining trees from selling to villagers etc. at places and then you can use that to get more items which gets you better until you can [very slight pause] say if it's like a zombie apocalypse you fight the boss. When we're playing the game we have to run around. Our game is physical not virtual. We run around and do stuff like when you're mining a tree you don't actually have an axe but you do have to find a tree . . .

I like performing. It's fun. It's not about being the centre of attention—it's just fun. People sometimes clap and applaud and say 'WOW!' and I like that. It makes me feel like I did a good job. I think I have good ideas and I like creating things because you can keep doing the same thing but then you can change it instead of doing the same thing—throw in something new—or you can reset everything to acquire something new, and it's fun to go back to the start and get to where you were again using that new item.

Please use my name because I want to be famous!

Xander's mother, Jacqui Dark, told me more about his unique personality:

Xander's interests are karate, gymnastics, swimming, chess, piano, singing, medical, science and mathematics, coding, debating, reading, building and creating.

His functioning short-term memory is excruciatingly bad. Xander can't get himself dressed and fed in the morning without constant reminders—left to his own devices, his breakfast would sit uneaten and he'd still be in his PJs when it's time to leave. He just forgets to do things. Instructions have to be simple and only one or two at a time—a complex list of tasks and he's lost. Getting out the door is excruciating—there are always seven things he remembers he needs just as we get out and he has to go back for them.

The joy is his incredible creativity. He can make up stories and games that charm and entrance adults and kids alike. His energy is boundless (blessing and curse) and he would never stop talking unless focused on another task. He asks endless questions—his curiosity is insatiable. He keeps coming up with thought experiments about the universe, but the constant, relentless demand for conversation and information is exhausting. When he's interested in something, he dives right into it. He'll look it up on his iPad independently and study it in depth. He then talks about it with huge enthusiasm (talks animatedly and loudly!). When he feels he's discovered something exciting he has to tell everyone about it and see them share his enthusiasm.

He draws energy from being around his friends. He LOVES people and interacting, and lockdown was hell for him. He treasures his friends and craves their company.

I suspect I have undiagnosed ADHD.

Jacqui expended her extraordinary energy on stage as an opera singer, while also following a huge range of other interests. A former teacher, she still set physics exams while actively engaging with friends who seem to number in the thousands.

If we return to our small foraging band in prehistory: Xander's hyperactivity would be incredibly valuable in searching out food and shelter. He is constantly distracted, but that also means that he is constantly aware of everything happening in his environment, including creatures much more deadly than Chipsy. With regard to the oral knowledge system, Xander's love of performing, his endless energy, his love of attention and his willingness to add novelty whenever he can would be superb traits for storytelling. His ability to bring his friends along with him in the performance he has created and his willingness to take risks intellectually and physically would make him a gifted member of the band. As a community, we don't want an entire population of Xanders, but we need at least one.

Dyslexia grants extraordinary spatial skills

People diagnosed with dyslexia have difficulty with reading, writing, spelling, memory, speaking and listening. From the Greek, *dyslexia* means 'trouble with words'. Estimates of its incidence vary hugely, as does the degree of disability, but somewhere between 5 and 20 per cent of all schoolchildren meet the criteria for a diagnosis of dyslexia. There is no evidence that children grow out of this 'learning disability'. Contrary to common understanding, in many cases, if not most, dyslexia is not just the reversing of letters and words.[11]

Recently it was announced that researchers have found 42 significant locations across the whole genome that are associated with dyslexia. It is up to 70 per cent heritable.[12]

There's no doubt that dyslexia is particularly painful because we live in a world dominated by the written word. Our education system places such a huge emphasis on reading that dyslexics are put at a disadvantage from a very young age. That would not have been the case through the vast majority of human history. Even since writing was invented, it is only in the last century or so that anyone other than the elites was expected to be able to read. Dyslexia exists as a 'disorder' only because of our modern world.

Those with dyslexia have demonstrable strengths in visuo-spatial skills, perception, exploration and out-of-the-box thinking. There is ample evidence of dyslexic strengths contributing to practical art skills, especially in three-dimensional visualisation. The value of these strengths in a non-literate world is substantial. It is obvious why evolution has kept dyslexia so prominently in the gene pool.

Without exception, everyone I spoke to with dyslexia described having been labelled as 'dumb'. They had suffered from extremely low self-esteem from their earliest school days. In no cases, from those still in school to the adults whose stories I include here, had their visuospatial skills been recognised and praised.

I am constantly irritated by media commentators and parents laughingly confessing that they never could do mathematics. Yet people who struggle with reading will do everything they can to mask the fact. Children are bombarded with messages that mathematics is hard and that reading must be easy because 'everyone can do it'.

When I first met Stephanie Houston, she lamented that she was unable to convince her husband, Marcus, that the astounding skills she could see in him were special. Having been told his whole life that he was dumb because he struggled to spell and his reading was slow, he refused to even consider her words. The more I heard of his abilities to visualise in three dimensions and then construct beautiful structures from any available materials (see Figure 6.1), the more astonished I became. I was horrified to realise the lifelong impact of the incredibly damaging words from his childhood.

Steph and I are strong at mathematics and science. We were told that we were smart, despite the fact that we both struggled with foreign languages and humanities, and found three-dimensional art concepts—not to mention actually constructing a viable building—inconceivable. We went to university and moved into the professions. Yet for the vast majority of human history, it was Marcus who would have been considered the smart one, as his skills would have been recognised as essential for survival.

Marcus describes his thought process:

If I want to make a sculpture, building or some form of brickwork, stonework or structure, I picture the entire inside/outside, what it will look like. I then use that picture

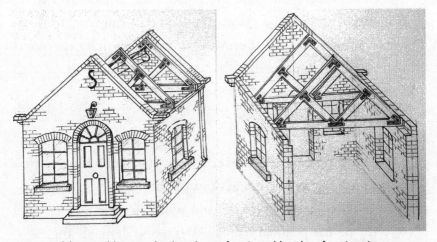

FIGURE 6.1 Marcus Houston's drawings, front and back, of a structure he will build for a client. These images were drawn from imagination of a building that did not yet exist. They took him about fifteen minutes.
IMAGES: MARCUS HOUSTON.

in my mind to work out what I need to complete each piece of that structure/task/project. Once I've got the image in my head, I finetune and work out measurements.

When I do a car up from just a rolling shell, I can see exactly what it will look like with the wheels on it, what colour, how it will sit once I change suspension, etc., how everything will mount together, and the finished product completed. Without measurements, I can swap the whole front end of a car, knowing where the bolts will go, etc., and get that image in my head of where it will all go, so I know where to start from.

I can imagine the finished product, even if it doesn't exist at all yet. I can then visualise all steps to get to that finished product.

In doing a landscape design or brickwork, I can see the whole end product, the whole backyard of what I'm going to create. Where the walls are, where the stonework and features are, where the plants will be, etc. I can then work out where the drainage needs to be, and everything else to make it all work properly.

Steph adds her view:

Marcus can visualise and make things out of anything and everything we have lying around the home. He is incredible at drawing and creating, and will draw me a picture of what he has planned to make next, and draw all angles in a lot of detail. Pages and pages of images so I can try to understand what he's doing. Everyone that comes to our house is amazed by the greenhouse entranceway he created, so safe, structurally sound and aesthetically pleasing purely from materials we already had! He does this time and time again with the brickwork, landscaping and stonework he does every day. It is incredible what he can do.

'Dumb' is not dumb. 'Dumb' is so often just different.

When I was at an art gallery with Jane Rusden, a successful dyslexic artist, we compared what we were seeing. A large three-dimensional installation of coloured plastic rectangles to me was simply a large installation of coloured plastic rectangles. To Jane, the colours were interacting with the spaces, including the negative spaces. She was seeing the whole and rotating it in her mind. I could not conceive of the way she was interacting with this artwork.

We looked at pictures of cave art. I saw the amazing animals; Jane saw amazing animals walking on the cracks of the rock and responding to the textures in the cave, representations of their behaviour on the ground above.

Jane explains her life as a dyslexic:

I vividly relive trauma from bullying and negative teachers over and over again. Years later, my nightmares are set in schools and having to complete my education. I still carry the trauma from my school days, where I was put down, seen as stupid and no good for anything, despite a private-school education with remedial classes from Grade 3 to Year 12.

I gained two degrees as a mature-age student, an honours degree in science and a visual arts degree. I put myself through hell and back to get them. I didn't dare tell anyone about my dyslexia, partly because of the negative treatment I was used to, and partly because I wanted to do it without extra help. But I still don't feel I'm good enough, even after all that study.

Jane often records her experiences directly, while in the bush (see Plate 6.1). She described one such painting:

I hadn't been there before so the first thing I did was wander around. I became innately aware of the topography. I put it all together in my head as a three-dimensional map. I'm always aware of what the topography feels like under my feet and imagining it in three dimensions.

When you're walking through looking at textures of leaves and textures of rock in the garden, you're going to get all sorts of plant structures and shapes and a lot of visual interest like that. It's very three-dimensional. I chose it because compositionally it leads your eye through a path.

You just know where you are in space. I have never been lost.

Many people with dyslexia are totally at home thinking in three dimensions. Their practical knowledge enables them to conceptualise structures created from whatever materials are available. That is a gift our prehistoric band of foragers would have valued highly.

People with heightened visuospatial skills can navigate well, enabling our band to explore new territories and return to camp safely. They can create art that responds to what they see and to the surfaces on which they can work. These are valuable skills needed by our band.

There is a glorious array of prehistoric art located in complex cave systems. The artists descended into chambers deep in the system, lit only by a flickering torch. If the torch should go out, their extraordinary visuospatial ability would have helped them to find their way out again; it was a matter of life and death.

Imagine the three diversities in our foraging band

It is speculative and oversimplified, I know, but just imagine our small band expanding their territory, while carefully retaining memory of what they already know. All members of the tribe have the advantages granted to them by having two copies of the knowledge gene.

They also have a few autistic individuals, who remain quietly on the sidelines, ensuring that the knowledge system built up over generations is retained accurately while being gradually adapted to incorporate new information. And the band has a few members with ADHD, who perform the knowledge and keep everyone entertained, re-enacting every novel experience in this new environment. The most important aspects of what they present will be appreciated by the whole tribe, including the majority, who we would describe as neurotypical. Slowly, they will be added to the formal knowledge system and included in the oral traditions so reliably repeated by those on the autism spectrum.

The people with dyslexia are great at building and navigating, but they are also stars when we think about enhancing the knowledge system. Who better to mark out the narrative pathways? Who better to decorate sacred places with art that takes advantage of the rock, wood and bodies they paint and the materials with which they sculpt?

The band thrives because of its neurodiverse members.

What about the full spectrum of genetically driven diversities?

There are so many genetically driven diversities that I don't have the space to include them all in this single chapter. Also, many of the diversities offer skills that may not be as valuable for Indigenous knowledge systems as autism, ADHD and dyslexia. It is also obvious that trying to find evolutionary advantages for every form of genetic diversity is both naive and doomed to failure. But I'd like to propose a few examples, necessarily speculative, in the hope that you might find it stimulating to contemplate why evolution has preserved these neurodiversities in the gene pool.

People with red–green colour blindness have been shown to be able to distinguish shades of khaki that those with normal vision cannot. The researchers hypothesised that this skill may have helped humans detect food items in complicated environments such as in foliage or grass.[13] People with myopia (short-sightedness) are able to detect tiny details beyond the visual skills of those with normal eyesight. Those with long-sightedness can take in broad landscapes with clarity. Both visual strengths would be valuable in noting the details of close and far environmental objects.

I have aphantasia, the inability to create visual images, which is estimated to affect about 4 per cent of the population. When people with aphantasia think of an apple or a sunset, they see nothing in their 'mind's eye', although they can describe both.[14] Research indicates that aphantasics live in the present, being unable to relive past events or simulate future ones. Consequently, they are less likely to suffer from post-traumatic stress or other mental illnesses that rely on visualisation.[15]

I have so little memory of past events that my husband says it isn't worth taking me anywhere because I will have forgotten it in a year's time. He is mostly joking . . . I think. My much-adored father's death at only 54 was devastating. I wept bitterly, but was soon over it and was able to support my mother, who never fully recovered. I have very little memory of his illness even though I nursed him. Unfortunately, I also have very little memory of him at all and am unable to visualise him, or my late mother. If I close my eyes and think of them, I see only grey mush.

I also have no 'mind's ear'. If I think of Beethoven's *Fifth Symphony*, I can only hear myself attempting to hum the notes. There is no orchestra and no actual sound. I cannot relive past traumas visually, or hear the unwelcome sounds. Any tragic event for our small band would leave many traumatised, but those with aphantasia might well recover faster and carry on.

Those with hyperphantasia are the reverse: they have extremely vivid mental imagery and recall past events with clarity. Our band would learn from their ability to recall the past accurately and to imagine the future.

There is evidence that blindness leads to an enhancement of other senses.[16] The genetics is complicated and linked to many genes.[17] And what about congenital deafness? Why has evolution left this diversity in the gene pool? Various researchers have mentioned that deaf individuals have a strength in pro-cessing and responding to information from their peripheral vision. This is likely fostered, it is suggested, because peripheral vision alerts them to movements in the environment for which hearing people depend on sound. Similarly, deaf people's ability to create, maintain and manipulate a visual image in their working memory is enhanced.[18]

I spent many hours with Michael Uniacke, fascinated as he shared his experiences as a deaf person and what he has learnt from many years working in deaf advocacy and writing about the experience of being deaf in a hearing world.

Four of the five children of Michael's hearing parents are deaf. Quoted in 2015, geneticist Rachel Burt explained that to date 88 genes had been discovered that, when damaged, can cause deafness with no other symptoms. It is likely that a damaged version of well-known gene GJB2 is responsible for as many as half of all cases of inherited deafness. Burt concludes: 'It may be that simply by chance, both Mike's parents carried a damaged copy of this (or another) deafness gene without knowing it.'[19]

Michael reflects on being deaf:

I began to commute on a road bike from Melbourne's inner northern suburbs, through peak-hour traffic, to where I worked in South Melbourne, a distance of 13.5 kilometres each way. Cycling in peak hour was hazardous. I was becoming annoyed with the rushing wind noise while on a bike, so off came the hearing aids. The transformation was immediate: I felt safer. There were two significant effects.

The first was to send my visual acuity and perception through the roof. I was thrown on raw survival, and became hyper-aware of everything around me as I cycled. That meant scanning the surrounds constantly, using peripheral vision, anticipating the behaviour of car drivers, and so on. I used a bike rear-vision mirror, made sure I was highly visible, and used other subtle tricks like making eye contact with car drivers next to me when stopped at the lights. The horrid wind noise was gone and it all felt cleaner and super-efficient.

The second effect was more subtle and personal. Without a hearing aid, I felt whole and complete. I was no longer a deaf-hearing hybrid needing a hearing aid to function. I was a deaf man, no more and no less; thrown back on myself I would deal with this as I judged best. I trusted myself to ride safely in the way that suited me best. Deafness became a way to complement and boost my other senses.

All the years I was growing up, hearing adults impressed on me the supreme importance of having on a hearing aid at all times. I was led to believe I could not function without it. But riding a bike in this way opened up the idea that I was perfectly fine just as I was, and in this way, deafness was not a defect. There were times when this very idea seemed quite shocking to me . . .

Michael and I met with a group of friends for lunch at a local cafe, and he described his experience when using his hearing aids:

It had been quite a few minutes since the five of us had ordered lunch. The waiter returned; he seemed anxious to check on something we had ordered.

At lightning speed and almost without effort, I made a series of observations from which I made deductions on what was about to happen, concluding that Lynne's order, and mine, were similar: variations on a Caesar salad. Therefore it was very plausible his query would be directed to the orders placed by Lynne and myself.

All that tuned me into certain keywords the waiter was very likely to use. My deductive reasoning helped me work out two words. It sounded like: 'Υπάρχουν anchovies στη sauce. Είναι εντάξει αυτό;'

I knew immediately that if he asked me, I would have to ask him to repeat. Out of a lifelong habit, I dropped my eyes to avoid eye contact. He then naturally addressed his query to Lynne. This gave me the additional advantage that by following what I heard of Lynne's response, 'Αυτό θα είναι fine', I could work out what the waiter had been asking. The sauce had an anchovy component and did we want that? That was easy. I smiled and said, 'No problem with me.'

This is an example of a process that I call 'rat-cunning', that as a deaf person I use to navigate and negotiate a world that is generally indifferent to the thousand communication complexities that deafness sets up. I don't have a ready definition of 'rat-cunning', but it bundles together components of deductive reasoning, the art of predicting what people are likely to say, environmental perception, past experience and educated guesswork. It all happens at lightning speed, at a very intuitive level.

Does Michael's idea of 'rat-cunning' offer an insight into what may have kept deafness in the gene pool?

Diversity in today's world

It was Albert Einstein who said, 'We cannot solve our problems with the same thinking we used to create them.'

What would a modern society be like if we took advantage of diversity rather than pressed for homogeneity? Power is so often granted to conservatives who benefit from conformity. History shows how much the pressure for conformity has led to discrimination. As a community, we lose the value of diverse thinking. We need new approaches to adapt to current crises, such as climate change, inequality and the prevalence of violence at the local, national and global levels.

Like many others, I wonder whether we would be much better off if the world adapted more to neurodiverse people, rather than always asking them to fit in. We need to take advantage of diverse strengths in contemporary society while remaining realistic about the challenges those living with these diversities face. Where are the autistic people, those with ADHD and dyslexia, the musicians and the artists, the deaf and the blind—indeed, where are all the differently abled—in the political sphere? It is time decisions aren't taken based only on their effect on neurotypicals.

We also need the neurodiverse to have access to education that will enhance their diversity-granted skills, and then we must listen to them. What the world needs now is neurodiversity.

CHAPTER 7

The prehistory of music through a knowledge gene lens

Having sidetracked from the research on the knowledge gene to explore the evolutionary questions with neuro-diversities, my focus returned to NF1. By this stage, I was convinced that the human NF1 allele was an essential driver for human knowledge systems. That meant I should be able to find evidence from our distant past that our musical and visuospatial skills are a fundamental contributor to our ability to store knowledge. So I headed off into the archaeology and anthropology of our species.

It was important that music, art, spatial and relevant cognitive skills were linked if the impact of a single gene is to be implicated in all of them. If music is very recent while spatial

155

abilities can be shown to date back 100,000 years, then the argument that NF1 is our knowledge gene becomes so weak as to be almost useless.

If this gene is in all of us, then I needed to show that the associated traits were universal, present in all human populations. They needed to have been so from our very distant past—which I am taking as from 126,000 years ago, as that is the earliest we can be sure that everyone had two copies of the human allele. I needed to show that they are innate, biologically driven, not just learnt by some. Our NF1 allele is not in our primate cousins, so the traits that support the claim that it is our knowledge gene have to be uniquely human. And critically, despite its fragility, there had to be a purpose important enough that evolution favoured it so strongly.

The first step was to explore the prehistory of music, the theme of this chapter. A critical component of my confidence was Vic's research, discussed in Chapter 1, which had shown that amusia affects, at most, 5 per cent of the normal population but 67 per cent of those with the NF1 disorder.[1]

I have very little musical ability but I do not have the severe musical impairment associated with amusia. My father tried to teach me to play the piano, but I couldn't hear wrong notes when I played 'Twinkle, Twinkle, Little Star' with one finger. At just eleven, I was instructed to mouth the words in mass singing at school because I put off those around me. In every class and in the final-year concerts, with everyone else singing wonderful harmonies, I was instructed to pretend. I will never forgive that music teacher. Her influence was so strong that I have never been able to sing, even softly, when anyone else is around. Not even 'Happy Birthday'.

For my daughter, Rebecca Heitbaum, music is a fundamental part of life. She plays oboe in orchestras and so has the experience I was denied. She offered her perspective:

> When everybody talks at once, it's just noise. If everybody plays at once, it's just noise. When you play your part adapting to those around you, you create music.
>
> You are constantly conscious of those around you. Someone might be playing a little fast, so you speed up. Someone might be playing a little flat, so you adjust your tuning to align with those around you. It's a feeling of community, of like-minded peers. I'm with my people; it's that strong an emotional connection.

I had written often about the value of music in terms of memory, but had not previously engaged with the evidence of the profound social impact of performing music in a group. The more I discovered about the social and emotional role of music, the more I cursed what that music teacher had stolen from me. Without music, human societies could not have bonded in ways essential for survival. One of the pillars of our evolution as a species is our musicality. As I discovered, the evidence was there both in the science and in the archaeology.

Is music a human universal?

I had one problem with the idea of musicality as a universal human trait, one that I spent many hours resolving. I found no shortage of references to the fact that music is found in every culture around the world—oral, literate, urban and rural.

If our musicality is driven by our knowledge gene, though, why does music around the world sound so different? Wouldn't we all be encoded to respond to, and therefore create, the same types of sounds? Yet Chinese music sounds so different from Western music, for example.

Eventually, I found the answer. Paleoanthropologist Dr Iain Morley was a lecturer in the School of Anthropology and Museum Ethnography at Oxford University. His work had greatly influenced my research on Indigenous knowledge systems and music from the outset.[2] Since those early days, Morley had published *The Prehistory of Music*, a book that would be a gift for my research.[3] He includes evidence from paleo-anthropology, archaeology, ethnomusicology, neuroscience, developmental and social psychology, and evolutionary biology, along with his own significant insights. His untimely death at only 46 was a huge loss to the field.

All our musical activity relies on the way we can voluntarily produce sequences of sound that we can make louder or softer, and higher or lower, within a defined rhythm. And, even better, we can do this in perfect synchronicity with one another.

Morley explains that it is the addition of an overall sound to a piece of music, its tonality, that gives the final clue. Primarily, although not exclusively, we tend to use a single scale for a piece of music. Morley describes the way all cultures divide the octave into five or seven discrete pitches, which are unequally spaced across the scale. It seems that, from infancy, as a consequence of our biology, we are set up to process unequal scale intervals better than equal ones. All cultures favour consonance and harmony. We don't like dissonance—that is, sounds combining tones that we consider to be clashing, and that lead to a harsh, unpleasant sound.

I could finally see the universality when Morley explained the 'perfect fifth' interval, which is ubiquitous because it sounds so good to all humans. The frequency of the first note in the scale is compared with that of the fifth note. You get a 'perfect fifth' when the two frequencies are in the ratio of 3:2. Morley concludes that our sense of harmony and dissonance, and our preference for the perfect fifth in our scales, are biologically driven common features, even though the musical behaviours built on these foundations vary hugely.[4]

Each note has a frequency. If you start at one frequency and then jump to the pitch that has double the frequency, you have jumped an octave. A musical scale is a sequence of frequencies, or tones, dividing up that octave. Western music is mostly based on the heptatonic scale—that is, it has seven tones. Chinese and Japanese music is often based on the pentatonic scale—that is, five tones within an octave, just like using only the black keys on a piano. The evidence suggests humans have been using the pentatonic scale extensively since well before the ancient Greeks.[5]

I will not divert here to explain further, as there's no shortage of music writers around who can do it so much better. Suffice it to say, I was now convinced that my concern about the universality was of no consequence.

Is human musicality ancient?

So music is universal for humans. But has it been so for as long as we have had a double dose of NF1 genes, at least 100,000 years?

The universality of music in all known human cultures is a strong indication that music dates back to human populations in Africa before they spread out all over the globe. The dates at which humans are thought to have first ventured out of Africa are constantly debated. The most recent research has taken that as far back as 86,000 years ago, by which time they were already in Asia. This date is based on fossils found in the Tam Pà Ling cave, in northern Laos.[6]

The ancestors of contemporary Australian Aboriginal cultures are now considered to have arrived in Australia at least 65,000 years ago.[7] From this point, they were separated from others of our species until well within the last thousand years. Music, dance and complex knowledge systems are known to date back tens of thousands of years in Australia. They had music.

Is musicality uniquely human?

If the human allele of NF1 is to be strongly implicated in our musicality, then music skills need to be *uniquely* human. But birds and whales sing, don't they? Do any animals synchronise and learn through music? If so, claiming NF1 as our knowledge gene would be questionable.

Michael Spitzer is a British musicologist and academic, author of *The Musical Human*.[8] Spitzer traced the biological capabilities for music—vibrations, rhythm and melody—back through the fossil record of animals for millions of years, long before our human ancestors walked the Earth. In a fascinating chapter on animals and music, Spitzer explains that music evolved separately in the 'musical' animals: insects, birds, whales and humans.

Insects are capable of perfect synchrony. Here, Spitzer introduced me to an invaluable word: entraining. This is the skill of counting or clapping or moving in perfect time with some external beat. And insects can do it. They will adjust their rhythm to entrain with another of their species in a way that no other group of animals can—except us.

Crickets can gather in choruses and chirp in perfect synchrony. But humans can take this skill further. We have what's known as metre, the ability to use recurring patterns and accents and growing beats, which we form into more complex patterns in ways no other species seems able to do.

Birds sing to attract a mate, repel rivals and establish a home. In some ways, their song has features of language, with reference and meaning, but they don't use syntax. Birds combine pitch, rhythm and contour in exquisitely complex structures. And their calls have distinct timbres.

The definition of timbre is often vague, but I found the best explanation to be that it is the character of a sound, the colour. Imagine a violinist playing the same note as an oboist and a trumpeter. Even at the same pitch, volume and length, you will be able to tell which instrument is which. The timbre is different. It is all to do with the waveforms generated by the individual instruments.

Only a small proportion of birds sing. Most birds have only calls, but a small proportion also have songs and can do the really clever stuff: vocal learning. That means they can modify sounds and acquire new sounds through imitation, something humans do with flair from a very young age.

Spitzer quotes some wonderful facts, such as that chickadees have only one song, while the brown thrasher, astonishingly, has 1800 different melodies. Chaffinches have twelve songs but they don't combine two or more of their songs to make more complex structures. Birds have amazing variety across the world, but we do the same thing as a single species.

Birds are much more sensitive to pitch, rhythm and timbre than we are, especially at perceiving perfect pitch. Father and juvenile king penguins can pick each other out in a colony of 40,000 breeding pairs, they are so finely attuned to the quality of their voices.

No wild birds have been observed keeping the beat of human music. However, a captive bird has been observed doing so. Snowball, a sulphur-crested cockatoo (*Cacatua galerita*), kept beat with music in situations where he could not be mimicking his owner.[9]

Living in an area to which sulphur-crested cockatoos are native, I can attest to their intelligence. At a previous house, the cockies, as we call them in Australia, would see me move in the kitchen, and would hang upside down from the verandah gutter, stare at me through the window and screech until I put out some seed. It is thought that parrots, along with a few other species, may be able to entrain to some degree because they are capable of vocal learning.

The ultimate avian vocal learner lived in the utopian garden of my childhood home: the Australian lyrebird. We learnt to stop worrying when the pump seemed to turn on spontan-eously: the mimics were at play. Spitzer tells of a lyrebird that could sing fragments of a Scottish folk song in the timbre of a flute, having learnt it from a flautist. This song spread through its feathered family for generations.

However, with only rare exceptions, bird songs stay the

same for generation after generation. Birds do not learn new songs. They cannot store knowledge in song and they cannot synchronise with each other.

Then we come to the greatest wild songsters of all: the whales. At a given place and time, all the humpback whales will be singing the same song. Whale watchers track the way the song changes year on year in melodies that are longer and richer than those of birds. But they don't store their songs, and every five years or so they seem to start a new cycle. And whales don't entrain.

Spitzer concludes that insects have rhythm and entrainment. Birds have a song. Whales have cycles of song and composition. It is only humans who build up a vast repertoire of songs and synchronise their rhythms in performance. And it is this trait that enabled humans to use music as an integral component of their knowledge systems.

It seems that our closest relations, the apes, don't use music at all. Gorillas, chimpanzees and bonobos, being closely related, share the same narrow range of very similar calls. Geladas and gibbons communicate in a seemingly musical nature, using rhythm and melody, synchronising (but not entraining) and taking turns. The duetting male and female gibbons may repeat their melodious call sequence many times during a single song bout, but this isn't considered music.[10]

There is currently no evidence that other primates are capable of entrainment, an integral part of human participation in music. Gorillas and chimpanzees can make rhythmic manual drumming when they're signalling, so they are physically capable of entrainment, but they do not synchronise their rhythms with each other.[11]

So we have established that human music is universal, ancient and uniquely human. But is it innate in us? That is, can we be sure that it is biologically driven? Without that evidence, the claim that the NF1 gene is of huge significance in our musicality becomes weak.

Is our musicality innate?

Morley argues that it is only through evolutionary processes that elements that are possessed at birth can arise. Babies are born with the perception of frequency, timing and timbre. They can understand some information from rhythm and tone alone, long before they have any language.[12] The logical conclusion is that at some point in our evolutionary past, musical behaviours offered selective advantages. Some people don't seem to think so.

Let us deal with the big-name doubters at the outset. And names in the study of evolution don't get any bigger than that of Charles Darwin. In *The Descent of Man* (1871), he wrote:

> As neither the enjoyment nor the capacity of producing musical notes are faculties of the least use to man in reference to his daily habits of life, they must be ranked amongst the most mysterious with which he is endowed. They are present, though in a very rude condition, in men of all races, even the most savage; but so different is the taste of the several races, that our music gives no pleasure to savages, and their music is to us in most cases hideous and unmeaning.[13]

At least Darwin acknowledges the universality of music. The rest I think we can dismiss.

Evolutionary psychologist Steven Pinker has been widely admonished for the opinion he expresses in *How the Mind Works* (2009).[14] Having concluded that music confers no survival advantage, he asks where it comes from and why it works. He suspects that music is 'auditory cheesecake': no more than 'an exquisite confection' that is a lovely side effect of the development of language. I strongly disagree. As music is a fundamental skill intricately linked to our ability to store information and bond us within our social groups, it is a tad more than auditory cheesecake.

Having noted that we naturally combine movement with music, neurologist and author Oliver Sacks describes this movement, this response to rhythm, as appearing spontaneously in human children before they have language. This trait has never been observed in any other primate.[15]

I find this fascinating. It brings back a long past memory from when my daughter was a baby, still unable to sit unaided. I noticed her spontaneously rocking for a few brief moments and then stopping. Some minutes later, she briefly rocked again. It worried me until I realised that it was when music played, such as the theme music to the news or the weather report. Ours was not a house where music was played very often, I regret to say, but despite this Rebecca became an accomplished musician. Her response to music was not because of her environment, nor dependent on language. It was innate.

Why do we have music at all?

But there's another question, one that, in my mind, has capital letters: the Big Question. Why does evolution consider music so important? And further, why do we need *two* communication systems, language and music? Does music offer anything that language alone does not?

I am not the first to ask the Big Question, and I am certainly not the most qualified to answer it. One who is eminently qualified to do so is Alan Harvey, neuroscientist and emeritus professor at the University of Western Australia. His book, *Music, Evolution, and the Harmony of Souls*, presents robust arguments supported by extraordinarily thorough academic research.[16]

Harvey devotes a lot of attention to the Big Question, looking for an explanation why, given the seminal importance of language, music is a second universal communication form. He provides the brain science to justify his conclusions. He also details the evolution of language, including reference to the language gene FOXP2.

My initial response to the Big Question is that language alone is simply not memorable enough. It lacks the enormously effective mnemonic impact that chanting or singing knowledge offers. Indigenous cultures depend on their living libraries, which are stored in music. Harvey alludes to this, but puts greater emphasis on the social and emotional benefits, which I now believe that at first I greatly underestimated.

As I imagined bands of prehistoric people moving across Africa, leaving it and settling in almost every habitat on Earth, Harvey's words really struck a chord. Creating strong bonds within each of those groups was a matter of survival. Throughout his book, Harvey returns to the idea that music has the ability to draw strong emotions even when there is no language involved. Music can communicate emotion, so critical when forming bonds between people. He details the huge range of ways in which music can elicit emotional responses both in listeners and in those performing.

Essential to the memory systems I talk about is how music can mimic the environment, but Harvey takes it further, talking about the way environmental sounds can generate appropriate responses because of our relationship with birds, other animals and every aspect of our environment. He explores how so much of our musical behaviour is shared with others, interpersonal interactions that he claims are central to our being.

These interactions will be familiar to all those who sing in choirs, play in bands or orchestras, or get up and move at concerts or festivals. They will recognise the impact of being an individual who becomes a contributing participant, something that Harvey sees as invaluable in social bonding.

When we talk, we take turns, but when we create music together, we are doing exactly that: making music *together*. Without social cooperation, humans could not have adapted to every environment on the planet.

Harvey gives the example of military music, which serves as a bonding force as troops march into battle. It inspires individuals to fight for the team. This drumming and movement and synchronised responses can bring on ritualistic, trance-like states, reducing fear and the perception of risk, leading to a willingness to sacrifice for the perceived common good. We see the same with sports teams and their fans, singing the team song or chanting in support of favourite players.

We also respond physically to music, especially when we are stirred to move; there are well-known impacts on our heart rate, temperature and that gorgeous tingling on our skins.

But Harvey also talks about the neurochemicals that are released in our brains and that are related to the formation of social bonds, the reduction of anxiety, and feelings of reward and gratification. The research on brain neurochemistry shows that there are close links between music, empathy, altruism and cooperative behaviours.

Morley agreed. His research into Indigenous societies demonstrates that musical activity is predominantly social and communal, which he considers was the same in Western society until recently.[17] This is an interesting point to ponder, especially as our pavements are full of individuals engrossed in their own musical worlds through earphones that block out their environment and isolate them from everyone around them.

Sacks goes further, imagining people dancing around the first fires, 100,000 years ago. He laments the loss of this primal role, as we now have a special class who do the composing and performing while the rest of us are passively listening. It takes a trip to a concert, music festival or church to experience music as a socially bonding exercise. That is not the daily experience that, Sacks considers, was our human interaction with music for tens of thousands of years.[18]

The fact that music began so deep in our evolutionary past possibly explains why it is one of the last of our skills to disappear when our cognition becomes damaged with age. Non-responsive in other circumstances, people with advanced dementia will sing along with familiar songs.[19] I'll return to this theme in Chapter 11.

It was intriguing when, during the Covid-19 lockdowns, ordinary people the world over emerged onto balconies and outside front doors to sing together, recordings of which were beamed around the world. Choirs and orchestras joined forces through social media, still determined to make music. Particularly moving was a video of the NHS Chorus-19, who rehearsed and performed virtually.[20] Over a thousand National Health Service workers in England joined the choir, which formed with the goal of lifting their spirits.

Is music just an extension of language?

Clearly, language and music both serve us well, but are they both part of the same evolutionary path? Or are they separate evolutionary gifts, which combine to give us so much more than either could alone?

Maybe music is just a flashier development of the prosody we use when we talk, the emphasis and intonation, highs and lows that give expression to what we say. Prosody can exist without language, such as when we cry, sigh, laugh, hum or chant a rhythm without words. Parents will often exaggerate their prosody when communicating with newborns, without the words being of any consequence. And when we do use words, it can be crucial to get the prosody right. A phrase such as 'You are so naughty' can be conveyed in very different ways depending on the prosody. This is a fundamental part of all human vocal expression.

If you have any doubts about how important prosody is in our communications, you only need to think about how often emails and texts are misinterpreted. When the language is separated from the intonation, so much of the communication is missing. That's why so many of us love exclamation marks and emojis! :-)

So could musicality be simply a side effect, beneficial though it is, of prosody?

The answer lies in considering rhythm, and in particular metre. Every culture on Earth produces a form of music with a regular beat that allows synchronised coordination between the performers and elicits a synchronised response from the listeners. The beat usually involves some notes being perceived as stronger than others in a regular pulse.

Normal language does not do this. There is not a regular pulse, nor regularly stressed syllables. We are so attuned to metre that research shows listeners usually tap or move slightly ahead of the beat; we anticipate and synchronise. This is unique to music and cannot be understood as a by-product of language.[21]

One of the world authorities on amusia, Dr Isabelle Peretz of the University of Montreal, has found compelling evidence that the capacity for music draws on multiple components within the brain. She reports on research on a group of individuals with congenital amusia, meaning they have been fully tone deaf from birth. Critically, they were still capable of recognising intonation and prosody in spoken language.[22]

Research led by Professor Willi Steinke of Queens University in Kingston, also in Canada, concluded that the melody and the lyrics of a song are actually stored separately in the brain. The brain then connects them so one will recall the other. Sing a tune and you'll recall the lyrics; hear the words and you'll recall the melody. The neural links are then reinforced every time you hear the song.[23]

The neuroscience is well beyond the scope of this book, regrettably, but the evidence points to the fact that language and music are independent domains, so magnificently entwined in our brains. But let's quickly turn to the genetic evidence.

As we saw in Chapter 2, the FOXP2 gene is essential for human speech and language. Those with a rogue FOXP2 display difficulty producing sequences of sounds, syllables and words, a problem known as apraxia. The NF1 gene is similarly specifically implicated in music and spatial skills. Those with a rogue NF1 gene are extremely likely to struggle with music, due to their amusia. There are without doubt other genes involved.[24]

Peretz and colleagues explored the possibility that FOXP2 also impacts musicality. They reiterated that congenital amusia compromises the normal development of musical abilities in between 1.5 and 4 per cent of the general population, with a substantial genetic contribution. Although amusia bears similarities to neurodevelopmental disorders of language, they found that having variants in FOXP2 was not likely to cause amusia.[25]

Given that the massive human genome has only been mapped for a few decades, and that there are around 3 billion base pairs of DNA, this research is in its infancy. But it is interesting to note that NF1 isn't the only genetic disorder that can impact music. As mentioned in Chapter 5, those with Williams syndrome tend to go the other way.

Williams syndrome is caused by a loss of genetic material from a specific region of chromosome 7. Like those with NF1, people with Williams syndrome often have difficulty with visuospatial tasks, but they tend to do well with spoken language and music. Oliver Sacks talks about Gloria Lehnhoff, a woman with Williams syndrome. Despite having an IQ below 60, Lehnhoff comprehends the thousands of arias in 35 languages that she can sing from memory. Although this is an extreme ability, even with Williams syndrome, virtually all those with the disorder are extraordinarily emotionally responsive to music.[26]

Rhythm rules

At the very heart of music is rhythm. It's actually fairly astounding how accurate our memories are for rhythm. The Renaissance scientist Galileo needed to time the descent of objects as they rolled down inclined planes. He hummed tunes, and in this way gained results with an accuracy beyond any mechanical timepiece available early in the 17th century.[27]

Galileo also used his heartbeat as a timing device. There is a popular misconception that your heartbeat tries to match the pulse of music you're listening to. It turns out that the pulse rates for music and human beings overlap significantly. Music is usually between 40 and 160 beats per minute, while the human pulse rate ranges from about 60 up to 150 beats per minute. Moving is what drives the heart rate up, not the music itself.[28]

In Chapter 3, I mentioned how the Haya, in north-western Tanzania, sing every phase of iron working to get the rhythms of the process right. Similarly, Scottish folk songs known as 'waulking songs' are sung in Gaelic by women beating newly woven cloth against a surface. This creates the felt-like texture and shrinks the fabric to increase its water-repelling properties. As the cloth grows softer, the tempo, or speed of the music, increases. It is mesmerising to watch.[29] Similarly, sailors sing their famous sea shanties as work songs, coordinating their physical hauling of ropes and raising of sails. These have become so enticing that they are often sung without a ship in sight.

My daughter, Rebecca, talked about a rehearsal when the source of the orchestra's pulse was absent:

> All rhythm has a pulse—it makes you feel good. The percussion provides that pulse. It keeps you on target, in time with each other.
>
> At one orchestra rehearsal without the percussion section, it just felt incomplete. There was a hole, something was missing. You didn't have that primordial beat. It was much harder to feel the music because the percussion is a relatively stable presence that you lock onto, whereas without that you don't have a constant. As an oboist, I adapt in pitch, and to some extent in tempo, to other sections. If I'm doing a duet with the flute, we stick together. But for the main la-la-la bits we stick with the boom-boom-boom of the percussion.
>
> Technically, everyone watches the conductor.

My Indigenous co-author on *Songlines*, Margo Neale, constantly emphasises that Indigenous music is part of an embodied knowledge system. You don't sit still and just listen. Gestures with hands and arms, rhythmic movements of the feet, a body bending and flexing—all these make music more memorable and enhance the social and emotional connections.

Margo also talks about the performances on Country as 'singing up' Country, as keeping it alive and as maintaining every connection: spiritually, intellectually and bodily. The songs represent not only the knowledge encoded but also the topography of Country. As the cadences rise and fall, so does Country. Enacting knowledge in dance increases recall; the message is conveyed not only in words, but also through every aspect of the movement.

I took Margo's advice. When encoding the large and confusing sandpiper family into my memory system for the birds of Victoria, I added movement. The little sanderling is a bird that moves in groups, in and out with the wavelets. On the sand at the edge of the tide, they are performing what appears to be the cutest of dances. In fact, it's because they feed on tiny crustaceans that are brought in by the waves and deposited in the wet sand, but for me they dance. Watching them was one of the highlights of my birding life. When I get to the sanderling when singing my birds, I dance in and out on my imagined beach. I have watched Indigenous cultures dance the behaviours of birds in a way that no writer could ever convey so perfectly. My sandpiper dance evokes that distant beach and brings it back to life.

It is also worth mentioning that neurogenesis—the creation of new neurons, enhancing the brain's plasticity—is enhanced by exercise. In particular, physical activity increases neurogenesis in what is known as the dentate gyrus of the hippocampus, an area of the brain that is important for learning and memory.[30] If you perform knowledge, you greatly increase your chances of remembering it. Your brain will be at its most receptive when you are dancing around, engaging physically with information. That is something worth considering when we ask school students to sit still for hours in class.

Morley's research with four hunter-gatherer cultures showed that they all used music in social events; it is always communal and almost always accompanied by rhythmic dancing. Important-ly for identifying music in pre-history, Morley also noted that the use of melodic instruments is minimal. Percussion and voice were all that was needed, and even the percussion was usually provided by clapping, slapping and foot stamping. These forms of accompaniment don't leave much around for archaeologists to dig up. With his usual wisdom, Morley warns that the earliest archaeological evidence for musical instru-ments cannot be assumed to represent the start of musicality in humans.[31]

To retain knowledge in the long term, repetition is critical. Much as we tend to hate repetition in normal speech, we are absolutely fine with it in music. That was something Georg Friedrich Handel knew well when he wrote the 'Hallelujah Chorus' in his *Messiah*.

Hallelujah! Hallelujah!
Hallelujah! Hallelujah!
Hal—le—lu—jah!
Hallelujah! Hallelujah!
Hallelujah! Hallelujah!
Hal—le—lu—jah!

Then, five times over:

For the Lord God
Omnipotent reigneth,
Hallelujah! Hallelujah!
Hallelujah! Hallelujah!

Try forcing someone to sit still for the 'Hallelujah Chorus' as you recite the words and they will not thank you. Sing it to them, preferably with a full choir and orchestra, and the effect is mesmerising.

How far back can we trace human musicality?

So we now know that music is universal, ancient, uniquely human, innate, heavily dependent on rhythm and usually embodied. As we saw in Chapter 3, there is no doubt that music is a fundamental repository of knowledge, a critical survival mechanism. Along with the way music bonds groups of people and impacts emotions, evolution had very good reasons for making sure the knowledge gene survived.

What happens if we look at the archaeological record through this lens of understanding?

Archaeology only records signs of musical instruments or acoustic resonance. Yet even that is enough to show how clearly music has been a key human activity throughout our prehistory. When we find archaeological evidence of flutes and drums and stringed instruments, we know that these are much later than the earliest examples, humans' first experimentations with music. The first musical instrument was surely the voice, accompanied by clapping and stamping and dancing. It would be a further step to use outside materials to add percussion, and then even further to form materials into instruments.

As music is an integral part of ceremonial events, performance spaces should also be taken as evidence of music. I'll be looking at ancient performance spaces in more detail in Chapter 9. More subtle indicators might include something

PLATE 1.1 A screenshot from the Horizon documentary *My Amazing Twin* depicting Neil Pearson (left) and Adam Pearson (right). The identical twins have neurofibromatosis type 1 but they have vastly different physical and mental symptoms. SOURCE: JAMES NEWTON FILMS (CREATIVE COMMONS LICENCE CC BY-SA 3.0).

PLATE 1.2 The hand of a person with multiple neurofibromas.

PHOTO: JFVELASQUEZ FLORO (CREATIVE COMMONS LICENCE CC0 1.0).

PLATE 4.1 Anna's home as a memory location, as I imagined it before and after I had met her. The difference as a memory location is immense.

IMAGE: LYNNE KELLY, WITH PERMISSION FROM ANNA.

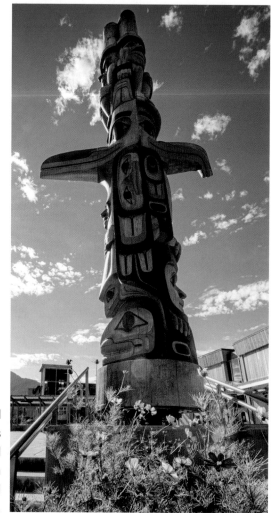

PLATE 4.2 The Sealaska Cultural Values Totem Pole represents the shared core values of all three tribes of south-eastern Alaska: Tlingit, Haida and Tsimshian. PHOTO COURTESY OF SHI.

PLATE 4.3 The Tsimshian clan house front was commissioned by the Sealaska Heritage Institute and represents the region's three tribes—Tlingit, Haida and Tsimshian. PHOTO COURTESY OF SHI.

PLATE 4.4 Tlingit group Sitka Kaagwaantaan Dancers performs at the Sealaska Heritage Celebration 2018. David Kanosh tells the origin story of the Kaagwaantaan, with guidance from clan leader Nels Lawson. PHOTO COURTESY OF SHI.

Plates 4.2–4.5 include copyright-protected clan property in the form of crests and regalia. These cannot be replicated without clan permission.

PLATE 4.5 Dancers of the Shangukeidí (Thunderbird) clan, also known as Chilkat. The Lukaax.ádi (Sockeye Salmon) clan, also known as Chilkoot, were invited to join the dance, along with the clan children. This photo contains sacred clan crests that are owned by the Shangukeidí and the Lukaax.ádi and cannot be replicated by those who are not members of these clans. PHOTO COURTESY OF SHI.

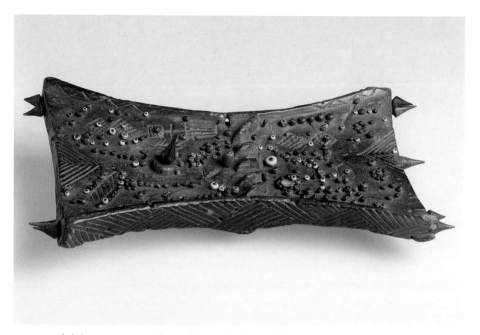

PLATE 4.6 A *lukasa* memory board from the late 19th or early 20th century and made from wood, metal and beads. PHOTO COURTESY OF BROOKLYN MUSEUM.

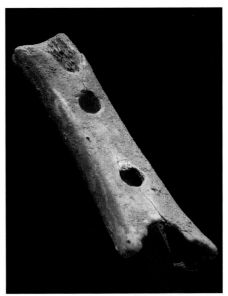

PLATE 6.1 Jane Rusden's nature journalling exercise, painted in a fern gully. IMAGE: JANE RUSDEN.

PLATE 7.1 The Divje Babe flute, measuring 11.4 cm long, alleged to be Neanderthal. National Museum of Slovenia. PHOTO: PETAR MILOŠEVIĆ (CREATIVE COMMONS LICENCE CC BY-SA 4.0).

PLATE 7.2 Various instruments from Grotte de la Roche, France, including a flute and decorated bullroarer. IMAGE: BASTIENM (CREATIVE COMMONS LICENCE CC BY-SA 3.0).

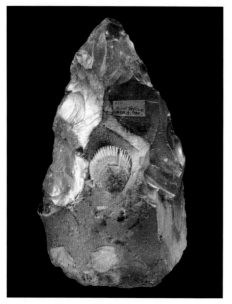

PLATE 8.1 Hand axe knapped around a fossil shell, West Tofts, Norfolk, England, *c.* 500,000–300,000 years ago. PHOTO: MUSEUM OF ARCHAEOLOGY AND ANTHROPOLOGY, UNIVERSITY OF CAMBRIDGE (CREATIVE COMMONS LICENCE CC BY-NC-ND 4.0).

PLATE 8.2 The Neanderthal engraving of a deer toe bone at Einhornhöhle. PHOTO: AXEL HINDEMITH (CREATIVE COMMONS CC BY-SA-3.0 DE).

PLATE 8.3 The Neanderthal structure dating back *c.* 175,000 years, deep in Bruniquel Cave, France. PHOTO: LUC-HENRI FAGE/SSAC (CREATIVE COMMONS LICENCE CC BY-SA 4.0).

PLATE 8.4 Lake Mungo, in south-western New South Wales, is no longer a lake.
PHOTO: DAMIAN KELLY.

PLATE 8.5 Nawarla Gabarnmang art site, Jawoyn Country, Australia.
PHOTO: JEAN-JACQUES DELANNOY (CREATIVE COMMONS LICENCE CC BY-SA 4.0).

PLATE 8.6 Female figurine from the Luba culture, West Africa, in the Royal Museum for Central Africa, Tervuren, Belgium. PHOTO: DADEROT (CREATIVE COMMONS LICENCE CC BY-SA 1.0).

PLATE 8.8 The oldest known figurine, *Löwenmensch*, from the German cave of Hohlenstein-Stadel (dated to 40,000–35,000 years ago). PHOTO: DAGMAR HOLLMANN (CREATIVE COMMONS LICENCE CC BY-SA 4.0).

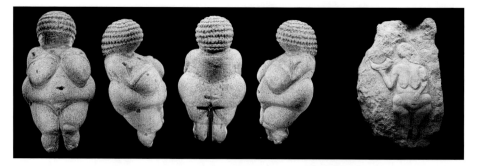

PLATE 8.7 Venus figures. Left: four sides of the Venus of Willendorf (dated to 25,000–30,000 years ago). Right: Venus of Laussel (dated to 20,000–18,000 years ago). PHOTOS: BJØRN CHRISTIAN TØRRISSEN (CREATIVE COMMONS LICENCE CC BY-SA 4.0) AND 120 (CREATIVE COMMONS LICENCE CC BY-SA 3.0).

PLATE 8.9 The Panel of the Horses from Pont d'Arc cave, an accurate copy of the Chauvet Cave. PHOTO: CLAUDE VALETTE (CREATIVE COMMONS LICENCE CC BY-SA 4.0).

PLATE 8.10 The polychrome ceiling at Altamira, Spain. PHOTO: MUSEO DE ALTAMIRA Y D. RODRÍGUEZ (CREATIVE COMMONS LICENCE CC BY-SA 3.0).

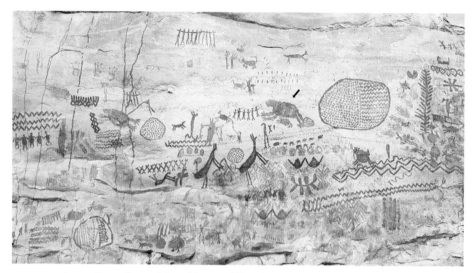

PLATE 8.11 A small section of the thirteen kilometres of the Serranía de La Lindosa rock art panel in Colombia. The arrow points to a proposed extinct giant ground sloth. PHOTO: JOSÉ IRIARTE ET AL. (CREATIVE COMMONS LICENCE CC BY-SA 4.0).

PLATE 8.12 The Sandhill Crane Site at the mouth of Currant Canyon, part of Nine Mile Canyon in Utah, United States. PHOTO: COURTESY COLORADO PLATEAU ARCHAEOLOGICAL ALLIANCE.

PLATE 8.13 The petroglyphs at Winnemucca Lake subbasin, Nevada, are North America's oldest. PHOTO: UNITED STATES GEOLOGICAL SURVEY.

PLATE 8.14 Petroglyph of a sleeping antelope, located at Tin Taghirt on the Tassili n'Ajjer, in southern Algeria. PHOTO: LINUS WOLF (CREATIVE COMMONS LICENCE CC BY-SA 3.0).

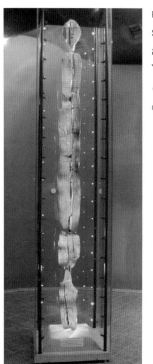

PLATE 9.1 The Shigir Idol, the oldest known wooden sculpture in the world, carved about 11,600 years ago. Sverdlovsk Regional Museum of Local Lore, Yekaterinburg, Russia. PHOTO: ВЛАДИСЛАВ ФАЛЬШИВОМОНЕТЧИК (CREATIVE COMMONS LICENCE CC BY-SA 3.0).

PLATE 9.2 Göbekli Tepe from a distance, and detail of a single pillar (above right).
PHOTOS: VOLKER HÖHFELD AND KLAUS-PETER SIMON (CREATIVE COMMONS LICENCE CC BY-SA 4.0).

PLATE 9.3 The Ring of Brodgar, with the Loch of Stenness in the background and the remains of the ditch in the foreground. PHOTO: STEVEKEIRETSU (CREATIVE COMMONS LICENCE CC BY-SA 3.0).

PLATE 9.4 The chamber in the Dwarfie Stane, on the Orkney island of Hoy, was carved out using only stone tools. PHOTO: DAMIAN KELLY.

PLATE 9.5 One of the 97 kerbstones at Newgrange, Ireland. The triskelion, the symbol used on the cover of this book, can be seen on the left of the stone. PHOTO: DAMIAN KELLY.

PLATE 9.6 Rujm el-Hiri, the 'Stonehenge of the Levant', Golan Heights, dating from the Early Bronze Age. PHOTO: SHII (CREATIVE COMMONS LICENCE CC BY-SA 3.0).

PLATE 9.7 Nazca jar (325–440 CE) with monkey. Held by the Brooklyn Museum, New York. PHOTO: LYNNE KELLY.

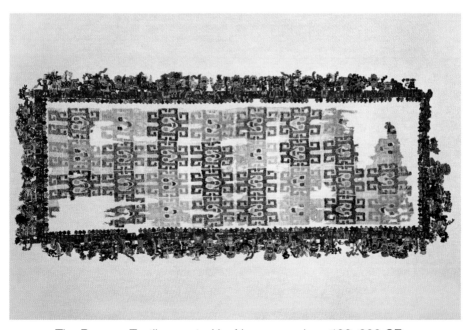

PLATE 9.8 The Paracas Textile, created by Nazca people *c.* 100–300 CE. Ninety figures decorate the intricate border. PHOTO: BROOKLYN MUSEUM, NEW YORK.

PLATE 9.9 Ahu Tongariki *moai*, platform and performance space, Rapa Nui.

PHOTO: PIOTR BIERNACKI (CREATIVE COMMONS LICENCE CC BY-SA 4.0).

PLATE 9.10 An artist's impression of Cahokia, 1100–1200 CE, with Woodhenge in the foreground. ILLUSTRATION: CAHOKIA MOUNDS STATE HISTORIC SITE, PAINTING BY LLOYD K. TOWNSEND.

PLATE 9.11 Pueblo Bonito: a digital reconstruction of the Great House in Chaco Canyon just before abandonment. ILLUSTRATION: COURTESY OF NASA.

PLATE 10.1 A section of a mural from the Tepantitla compound in the Mesoamerican ruins of Teotihuacan, Mexico. There are at least twenty speech scrolls in this detail. PHOTO: TESEUM (CREATIVE COMMONS LICENCE CC BY-SA 2.0).

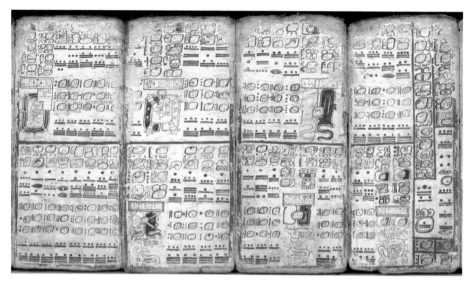

PLATE 10.2 Four sheets of the Maya Dresden Codex (pp. 55–58), depicting eclipses, multiplication tables and the flood (*c.* 1200 CE).

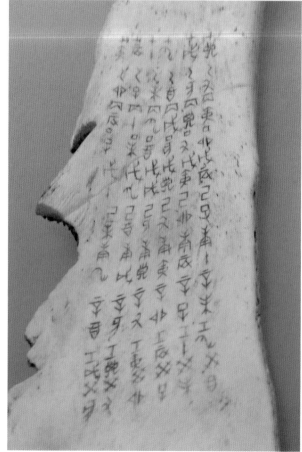

PLATE 10.3 Table of the Chinese sexagenary cycle, an ancient counting system, inscribed on an ox scapula, corresponding to the reigns of the last two kings of the Shang Dynasty. PHOTO: GARY LEE TODD (CREATIVE COMMONS LICENCE CC BY-SA 1.0).

PLATE 10.4 A portion of the main rock painting on Huashan Mountain. The paintings are located on a cliff face along the west bank of the Mingjiang River in Yaoda Town, Ningming County, Guangxi, China. PHOTO: ROLFMUELLER (CREATIVE COMMONS LICENCE CC BY-SA 3.0).

PLATE 10.5 A small segment of the handscroll *Along the River During the Qingming Festival* created by artist Zhang Zeduan (1085–1145), Northern Song Dynasty, and housed in the Palace Museum, Forbidden City, Beijing.

PLATE 10.6 Chinese artwork *Walking on a Mountain Path in Spring*, by Ma Yuan (1160–1225), in which calligraphy is an intrinsic component.

PLATE 10.7 A manuscript from the *Mewar Ramayaṇa*, depicting Rāma slaying Rāvaṇa. It was created by scribes and painters employed by the Kingdom of Mewar and dates to the 17th century. Held by the British Library.

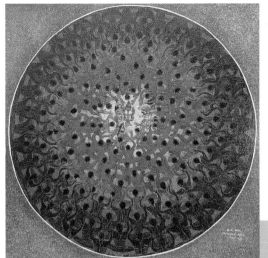

PLATE 10.8 A contemporary painting titled *Kecak (Monkey) Dance*. IMAGE: COURTESY OF REBECCA AND RUDI HEITBAUM.

PLATE 10.9 Cuneiform tablet: hymn to Marduk, possibly Babylonian, 1st millennium BCE. IMAGE: COURTESY OF THE METROPOLITAN MUSEUM OF ART.

PLATE 10.10 Part of the decoration of an Attic red-figure vase, depicting Greek mythological characters: Eurynome, Pothos, Hippodamia, Eros, Iaso and Asteria; *c.* 400 BCE.

PLATE 10.11 A leaf from a manuscript of Valerius Maximus from the workshop of Pierre Remiet, *c.* 1380–90. IMAGE: COURTESY OF THE METROPOLITAN MUSEUM OF ART.

PLATE 11.1 A small portion of the handscroll *Spring Morning in the Han Palace*, by Qiu Ying (*c.* 1494–1552), which measures 30.6 x 574.1 cm. Held in the National Palace Museum, Taipei.

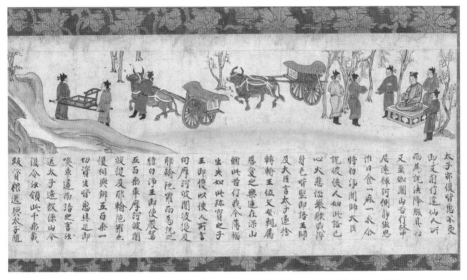

PLATE 11.2. Part of a Japanese narrative scroll, *The Illustrated Sutra of Past and Present Karma* (*Kako genzai inga kyō emaki*), by an unidentified artist, late 13th century. IMAGE: COURTESY OF THE METROPOLITAN MUSEUM OF ART.

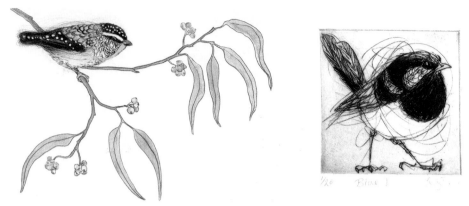

PLATE 11.3 An illustration (left) of a spotted pardalote to show accurate colours and habitat, and 'art for art's sake' (right)—a drawing of the superb fairywren that captures personality. IMAGES: BRIDGET FARMER.

PLATE 11.4 Jane Rusden's painting of a brown falcon conveys the power and presence of a bird that dominates the food chain. IMAGE: JANE RUSDEN.

as seemingly unrelated as a bundle of objects for which there is no obvious practical use. Some members of the Indigenous American Blackfoot Confederacy, for example, have specific songs for each act in the ceremony, and for each ceremony they have a bundle of objects that is opened to ensure that the correct songs are sung—each song is associated with one of the objects. Some of these bundles can contain over 160 items.[32]

Unfortunately, I cannot in this short chapter even skim the surface of the research into music making in the distant past, but I can present some of the most exciting finds. There is robust evidence that music appears very early in the archaeological record, at a time when we know that all humans had two copies of the knowledge gene. We will also see that music was part of a complex of activities using the other skills that are directly linked to this gene.

The first instruments

The earliest musical instrument we have evidence for is the Neanderthal flute, which long pre-dates any other such archaeological find. The claim, however, must remain provisional because there is significant debate over how to interpret the object. The juvenile cave bear femur with pierced spaced holes was found in the Divje Babe I cave in Slovenia. The archaeological context in which it was found shows no evidence of modern humans, only Neanderthals.[33] The Neanderthal flute has been dated to about 67,000 years ago, over 30,000 years before the first *Homo sapiens* wind instrument, although that dating is disputed.[34]

The Divje Babe flute is prominently displayed in the National Museum of Slovenia in Ljubljana, where it is labelled as a Neanderthal flute (see Plate 7.1). I really, really want them to be right, because Neanderthals carried two copies of NF1 and I would love to have concrete evidence of their musicality. I shall have to wait and hope.

Indigenous cultures tend to use hollow instruments—an example is the Australian didgeridoo. These are most often made of wood, so the archaeological record almost certainly only represents a small proportion of all the musical instruments used so long ago in human history. Bone flutes will survive, while wooden flutes will not.

While considering the knowledge gene influence as an intermeshed suite of skills, it is worth noting that the early flutes made by *Homo sapiens* are commonly found in decorated caves.[35] Creating such instruments requires spatial skills and a significant attention span.

Morley presents a very detailed analysis of the extensive archaeological record of the earliest forms of pipe.[36] He notes the importance of establishing that these pipes were deliberately made by humans, and that the perforations were not due to animals chewing on the bones.

In the Ach Valley in southern Germany, the pipes found in the caves of Geissenklösterle, Hohle Fels and Vogelherd are considered the most complete and oldest known. There are four pipes made from wing bones of large birds and four made from mammoth ivory. Morley notes that the ivory pipes would have taken a great deal more effort and precision work than those from bird bone, yet produce an almost identical instrument. He describes the technically complex and challenging tasks that would be required for the ivory pipes to produce sound: sawing, refitting and binding. Clearly, the material was significant; it is also evident that the technology was not new but a finely honed, sophisticated activity.

The Isturitz cave in the French Basque region of the Pyrenees is the richest source of evidence for intentionally produced sound-makers. When looking at this evidence through the lens of the knowledge gene, it is worth noting that a large chamber, the Salle d'Isturitz or Grande Salle, also features a stalagmite with amazing acoustic properties and decorated with images of animals. This combination of music and art is a recurring theme. Along with thousands of stone tools at Isturitz, there are engraved stone plaques and cave art. Some of the materials used, not being found locally, must have been traded or transported over distances up to 100 kilometres. Although not continuously used, this was a major gathering site for more than 25,000 years.

On two of the Isturitz bone pipes are incised lines, which Morley concludes could not have been merely decorative. He refers to research by Lawson and d'Errico, who consider the marking sequences to be similar to those that they refer to as 'Artificial Memory Systems', through which knowledge is conveyed.[37] Having been impressed by d'Errico's earlier work on artificial memory systems from the Upper Palaeolithic, I was delighted to see that knowledge systems were being included in interpretations of this cave and its wonderful collection of diverse artefacts.[38] Isturitz has all the hallmarks of being a phenomenal knowledge space.

Other objects have been used to make noise, too. Toe bones, almost always from reindeer, were used throughout the Palaeolithic as whistles. Technically, these are called *phalangeal whistles*. It is disputed whether they are musical instruments or simply whistles used to attract attention or communicate, as their sound travels over great distances.

Equally ambiguous in purpose are bullroarers. These are flat pieces of various materials, including wood and stone. Spun on the end of a cord, they create sounds varying from low rumbles to high-pitched screams. They are used throughout the world by Indigenous cultures and also appear in the Palaeolithic record. In Australian Indigenous contexts, bullroarers are often decorated with geometric designs that represent knowledge of navigational paths and campsites, animal tracks and plant species along with symbols of identity.[39]

As wood seems to be the most common material, it is likely that most bullroarers have not survived in the archaeological record. A bullroarer made from reindeer antler found in the upper Palaeolithic layers at the French cave site of Grotte de la Roche was covered with red ochre (see Plate 7.2). At 180 millimetres long and up to 40 millimetres wide, it was incised with lines (see Figure 7.1).[40] The combination of sound and symbols is a strong indicator that these objects were part of a knowledge package.

Pieces of wood, bone or stone with grooves cut in them have been interpreted as rasps, which create a staccato vibration when rubbed with another object. Many such objects are

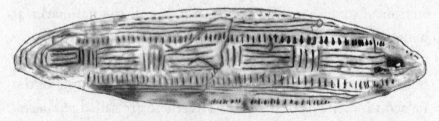

FIGURE 7.1 Markings on the Grotte de la Roche bullroarer. IMAGE: LYNNE KELLY.

known from around 19,000 years ago but whether they were used as musical objects is debated. They have also been interpreted as recording calendrical information.

But what about an ancient bow? Just because a bow is found does not mean an arrow is involved. There are many claims for the first bow and arrow, but the most robust dates back to around 60,000 years ago at Sibudu cave, in southern Africa.[41] The earliest European bow and arrow dates to 54,000 years ago from Grotte Mandrin, in Mediterranean France.[42] A wall image at Çatalhöyük, in Türkiye, dated to about 9400 years ago, shows a figure holding two bows with no arrows. Is that more likely to refer to music than hunting?[43]

The temptation to speculate is irresistible. What are the chances that the prehistoric hunter, hanging around with a taut string on his bow, didn't twang it? And, once he found that his twang made a note, what are the chances that, sitting there and waiting for his prey, he didn't move one thumb and finger together down the string and twang it again, discovering a different note. People appear to have done so later in prehistory. A cave painting dating to approximately 17,000 years ago at Trois-Frères, France, depicts a masked dancer playing a musical bow. In recorded history, musical bows have been played for thousands of years, including by people of Asian, Pacific, African, American and European cultures.[44]

Single-string musical instruments are still played today, such as the ektara, once the prized instrument of wandering bards and minstrels in India, and now used in modern-day Indian music as well as in Bangladesh and Pakistan.

Pythagoras used a monochord, a single string stretched across a sound box, to explore the mathematical ratios between musical intervals. When I taught the physics of music, my students con-structed monochords with the intention of copying Pythagoras's experiments. It was delightfully common in my classes for students to deviate significantly from my clear instructions and do their own thing. There was a lot of weird music created from this very simple instrument.

We know from Indigenous cultures all over the world that percussion is the primary musical accompaniment, if there is any, to voice and rhythm. We also know that most rhythm instruments, primarily drums of some sort, are made from wood and animal hides, of which there will be nothing left in the archaeological record. But every now and then something turns up that indicates how rhythmic percussive instruments were used in the very distant past.

The world abounds with naturally ringing stones, and there is ample evidence that humans have been very willing to take advantage of them. These natural stone features are sometimes referred to as 'rock gongs'. For example, in the Midlands of KwaZulu-Natal, South Africa, rock gongs were used by the San people to communicate over long distances.[45] They are still used by First Nations people as they seem to have been for thousands of years. They are often marked in some way, such as by cup marks in the Neolithic and Bronze ages.[46]

A beautiful example is sometimes, possibly ambitiously, known as a Palaeolithic 'orchestra'. There is convincing evidence that a set of six mammoth bones discovered in Mezin, north-eastern Ukraine, were deliberately and repeatedly struck, particularly in the areas that gave the highest resonance. These bones have been incised and painted. Piles of red and yellow ochre were found at the site. These ancient musicians used a shoulderblade, thighbone, jawbones and skull fragments that produced a range of tones when struck with beaters. To add to our prehistoric orchestra, there appear to be rattles made from ivory. The site consists of five dwellings, which could have held up to 60 inhabitants. Archaeologists concluded that one Mezin dwelling is non-residential, containing all the artefacts expected for seasonal feasts. And there is a figurine of a woman who appears to be dancing.[47]

At Mezin, we have the same combination of indicators that I have found all over the world in archaeological reports and in contemporary First Nations cultures: art, music and performance spaces. Too often the archaeological reports talk about the art, or about the musical instruments, or about the figurines, or about the space for performance, or about the ability to create three-dimensional tools. But the knowledge gene influences the entire suite of skills. Examining artefact types in isolation loses precious data that could be used to interpret what was really going on.

The Venus of Laussel is a rare example of a prehistoric bas-relief, dating to around 20,000 years ago. She was covered in ochre, and is seen holding an incised bison horn in a pose that looks as if she's playing it as a musical instrument.[48] She can be seen on the right of Plate 8.7.

There are many more indicators of music from our distant past for which there isn't space here. Then there are also others I will talk about as they appear in monuments, the theme of Chapter 9. For example, at Göbekli Tepe, Anatolia, Türkiye, occupied from about 11,500 years ago, the megaliths have been described as sounding stones. At Çatalhöyük, also in Türkiye, occupied from about 9400 years ago, there are wall paintings representing what appear to be musical instruments and ceremonies. In monuments all over the world, including those at Stonehenge and Malta, the impact of sound has been found to be hugely significant.

You may have noticed a lack of reference to Asian prehistory. The main reason is my great fascination with China, the only country where a script can be traced right through from early cave markings to a language still spoken, and to a script still written today. The prehistory of China will feature strongly in Chapter 10, where I look at what happens to the knowledge gene skills as the long pre-literate past gives way to writing. Literacy includes the first musical notations, the earliest of which date to about 5500 years ago.

Music in the caves

All good scientific hypotheses must be able to be falsified. The claim is that the knowledge gene was hugely significant in both music and art, and so we should be able to see these two traits constantly appearing together. Wherever literacy has dominated learning, visual arts and performance skills are pretty much separate domains. In First Nations cultures, music is intricately interwoven with art.

In his book *Dancing at the Dawn of Agriculture* (2003), archaeologist Yosef Garfinkel refers to dancing as a critical part of the communal performances that conveyed knowledge before writing.[49] He talks about how prominent dance is in the art during the Neolithic. Garfinkel analysed over 400 artefacts depicting dancing from 170 sites found in Asia and into Europe dating from 10,000 to 6000 years ago. He observed that circles of dancers are far more prevalent than those in a line or, even more rarely, a couple.

It is even more revealing that so many reports of caves and acoustics note that art, especially abstract symbols, appear more frequently at points of maximum acoustic resonance.[50] Reznikoff and Dauvois made a detailed study of the acoustic effects in three Palaeolithic caves in the French Pyrenees. They found that the art in the caves, dating to around 20,000 years ago, correlated highly with the points that resonated most strongly to particular notes. Most of these paintings were within a metre of a point of resonance, while most points of resonance had a painting. Often the art at these resonating locations was not easily visible or accessible. In some of the most difficult locations to reach, the art is only in the form of red dots, but it is there. These painting activities complemented sound-producing activities. The full potential of the resonance in these caves can only be elicited by the entire range of a human voice. Reznikoff and Dauvois suggest that voice and resonance were part of the way in which the poorly lit caves could be navigated.[51]

Using music as a navigation device could explain how these people could have regularly accessed such complex spaces safely. If their unreliable lamps failed, they would still have their location markers and be able to work their way out again through sound.

Calcite formations in caves have been covered with abstract designs too insignificant to be seen from a distance. They are found on features that produce clear tones when struck, leaving evidence of the impacts in the wear patterns. It is assumed that this was happening at the same time as the art, around 20,000 years ago. This combination of sound and art is found in many sites in France, Spain and Portugal.[52]

There is an abundant and ubiquitous use of music and spatial skills together. From as far back as we can trace human behaviour, the pervasive use of music and spatial skills abounds.

Our ancient musicality has implications for today

The more aware we are of why evolution has preserved our innate musicality, the more we can tap its potential. We can consciously take advantage of the ability for music to impact our mood. I have started taking note of music that inspires me, something many others I've talked to have already done. For some reason, my calligraphy is better if I'm playing Strauss. My control of watercolour seems to respond well to particular operatic arias. Certain popular songs lift my spirits and make me want to dance and move, which is obviously very good for my health. Other songs inspire me to think more deeply about past experiences and learn from them. I rarely had music playing around me in the past, but it's very much a part of my present.

I am now creating a music list to trigger particular moods and activities. I'm deliberately creating tunes that get stuck in my head and rewording them for what I want to remember. This is something I've seen work in schools; it's quite a common practice among those who have already discovered this invaluable trick of the trade.

Unlike rote repetition of information, which must be laboriously learnt, singing knowledge is something that you can do for the pleasure of it. I sing my bird families in the shower almost every night, because going into the shower triggers the song and makes me feel like singing it. What I've discovered is that I have been biologically encoded to take advantage of my musicality all my life. If only I'd known sooner.

I've been talking to people who are involved with community choirs. The same stories are told repeatedly. There are some whose musical skills are more akin to mine than to Pavarotti's. But they still turn up without fail, and they belong. I heard about one elderly lady who, despite mobility issues, always ensured that she got her dose of social connectedness at her community choir.

The threat of dementia is a major concern for older people. Dementia covers a range of disorders, which a significant proportion of the older community will never experience. However, it is a major cause of death and a distressing part of life for so many.

The evidence for the power of music with dementia patients is overwhelming. Given the ancient nature of the knowledge gene, this is not surprising.

Oliver Sacks writes that even in those with advanced dementia the response to music is preserved. He describes his astonishment when he sees individuals who are mute, isolated and confused, yet who still warm to music. They recognise it as familiar, and will start to sing and bond with others present. Non-reactive, non-responsive patients become alert and aware of what's going on. The music also calms some who are agitated. He observed groups, many of whom were virtually speechless, singing together and, for a while, bonding with others. Even if normally immobile, people with advanced dementia can rise and dance. Sacks considers the consciousness awoken to be perhaps the deepest of all. Only music has this impact.[53]

One twin study reported that musicians were 64 per cent less likely to develop mild cognitive impairment or dementia. This was reported in a paper identifying three studies that investigated the specific relationship between playing a musical instrument and the subsequent incidence of cognitive impairment and dementia. They all reported a large protective association.[54] I wish I'd known that before I gave up the piano.

Every one of us has two copies of the gene that offered humans the extraordinary gift of music to use as part of our ability to learn and to share our understanding. Humans have taken advantage of this prized skill throughout our long history. The more we engage with music in the heart of our lives now, the more we will be fully human.

CHAPTER 8

Prehistoric art and the knowledge gene

Over ten years ago, my research on Indigenous knowledge systems and archaeology took me back as far as the Neolithic. In 2015 I wrote in an academic monograph: 'Given that this book emphasises the link between the art of oral cultures and the knowledge system, it would be fascinating to explore how far back this link can be made, and how that relates to the origins of modern human thought.'[1] Despite my husband's fascination with human origins, too many other directions proved enticing, and this was a path I had not explored. The discovery of the mutation that led to the knowledge gene NF1 was exactly the catalyst I needed to search back as far as I could.

The prehistory of art is vivid and colourful. But is art purely beautiful, or is there a lot more to why humans have created glorious images for tens of thousands of years? Understanding the potential granted by our knowledge gene explains why abstract designs sit alongside representational rock art in huge panels. It explains why, as we saw in Chapter 5, only humans can take an object such a rock or lump of wood and visualise a completely different shape hidden within it.

The magnificent horse heads and glorious bison in cave art are carefully selected segments of the galleries that appeal to modern sensibilities. Very often they are just a small part of large panels consisting of all sorts of representational and abstract motifs. We need to consider art in its context.

At the beginning of Chapter 7, I identified five aspects of music that had to be shown as valid for me to claim that music was a biologically driven skill. To consider the same five criteria for art is a lot easier, as the archaeological record stores evidence of art a lot better than it does music.

Is art universal, ancient, innate, uniquely human and incredibly valuable?

One of the leading writers in the field of rock art, Paul G. Bahn, has demonstrated that personal adornments, art and imagery occurred at archaeological sites widely separated both in space and time. As he convincingly argues, humans would not have suddenly become artists. The skills needed to create both abstract and representational images must have gradually built on the abilities of many previous generations. Bahn titles an entire chapter of *Images of the Ice Age* (2016) as 'A Worldwide Phenomenon'.[2]

Bahn's view is widely held among archaeologists: human art is universal and ancient. Next is innate.

Given the material to draw, all children will do so. Independent of their culture, they will all pass through stages by which their creations become increasingly meaningful and representative.[3] They don't need to be taught. The desire to draw is innate.

This is something I witnessed when friends brought their baby to visit. Barely crawling, she still managed to find a pen and draw with vigour. She scrawled, unnoticed, for long enough to cover a significant area of our freshly painted wall. She had never held a pen before, her parents not being aware she was ready to start her artistic journey. I wished she had waited a day longer to display her innate, genetically driven skill.

No other living species spontaneously makes images on every surface they can find. No other species creates sculptures, tiny and huge, portable and fixed in place. As it is uniquely human, art is one of the defining characteristics of our species.[4]

Finally, did art, in all its glorious forms, serve a purpose so valuable that evolution favoured it? From the discussion in Chapter 4, obviously I think that it did. But it is through looking at the archaeological record that it becomes clear that art has served as a knowledge system for a very long time. And that is the theme of the rest of this chapter.

In the last chapter, I quoted neuroscientist Alan Harvey asking why we needed two communication systems, language and music. Along with the social and emotional impact, music can store information in a far more effective way than language alone. Now I'm going the change the question and ask why we need *three* communication systems. Is it because, in combination, music and art enhance the ability of language for communication? In the next chapter, I'll add the connection to large spaces. All these systems combine to form the most effective knowledge system imaginable.

Usually we think of music and art as separate domains. We think about memory and science but not when we're thinking about music or art. The time has come to pull down those silos.

What is art?

Australian archaeologist Iain Davidson is an emeritus professor of archaeology and paleoanthropology at the University of New England. I will use his very straightforward definition of art: 'the making and marking of surfaces'.[5] So cut marks on bones caused by removing flesh are not art, but lines etched onto the same bones for the purpose of leaving marks are art. There are many definitions and debates out there in the literature, but that is what I am going with.

Generally, art on wood or in sand can't survive in the archaeological record. What archaeologists find will only be a tiny portion of the full artistic expression of our ancient ancestors.

Could the ancient markings, however, just be 'art for art's sake'? Possibly, but the only insight available to us is from oral cultures that have long created rock art and still do. My co-author on *Songlines*, Margo Neale, is an expert on Indigenous art, both that from her own heritage and from all over the Australian continent. I asked her opinion. She wrote:

> I doubt that Aboriginal visualisations would be for aesthetic purposes only. This is not to say that the odd individual may not indulge themselves. It is well known and stated by many

an anthropologist that aesthetics are not the primary motivation for the production of 'art'.

Being a non-text based society, 'art', as it is called now,
was to communicate in the absence of text, and to transmit
knowledges. This is not to say aesthetics did not have a role,
because in a multi-sensory experiential learning system the
images have to activate all, or many, of the senses for the
knowledge to be absorbed. It has to activate various parts of
the brain. Aesthetics may be a vehicle for transmission but it
is neither the reason nor the content.

I often think of the sense of shimmer that some good
bark paintings from Arnhem Land have. This is achieved
through fine crosshatching or fine dotting, often in all
white. It represents, in some cases at least, the appearance
of the flickering light on painted up bodies in motion for
ceremony at night. They say the greater the shimmer in the
work the greater the spiritual power.

I also asked David Kanosh, the Tlingit Elder and storyteller
in Alaska, for his view. He replied: 'Indigenous people do not
put a lot of time and effort into something which serves no
practical purpose. It's more than just pretty art.'

From a third continent, in the Democratic Republic of the
Congo, the Luba people are famous for their elegant carving,
which I consider to be some of the most beautiful sculpture
in the world. I was very fortunate to have a brief exchange of
emails with the late Professor Mary 'Polly' Nooter Roberts,
who introduced me to Luba art.[6] Polly and her husband, Allen,
worked closely with the Luba people and wrote extensively
about their artworks. Polly explained: 'They are objects of

aesthetic brilliance and achievement made by artists, but they were made for other purposes: for education, or healing, or governance, or spiritual mediation. I always say they are more than art.'[7]

Indigenous archaeologists and those working with First Nations cultures recognise this. In the Australian context, Laura Rademaker and colleagues write: 'The stunning galleries of art, curated and preserved in rock shelters or across plateaus, are therefore also archives . . . The landscape holds the songlines and stories of the continent. Rock art simply makes this deeper record visible.'[8]

When we look at ancient cave art for which we have no continuous culture, we cannot possibly know the interpretation. We can surmise that knowledge was represented and possibly the genres of knowledge, but not much more. I reluctantly suggest that the search for absolute meaning is fruitless.

Objects in archaeological reports, including art, are usually divided into 'utilitarian' or 'non-utilitarian'. Most objects considered to be art are assigned to the latter category. The word *utilitarian* is defined as 'designed to be useful or practical rather than attractive'.[9] I believe that almost all ancient art, like that of contemporary oral cultures, is utilitarian. That doesn't mean it can't also be beautiful; indeed, it's better if it is. Just as a modern textbook is designed to look as appealing as possible, its primary purpose is to convey useful information: it is utilitarian.

When did art start?

Tantalising hints of our knowledge gene can be found with a really early human species. *Homo erectus* ('upright man') first appears in the archaeological record from about 2 million years ago. Although there are earlier examples of very simple tool-making, *H. erectus* is credited with the first 'bifacial' tools—those with two finely chiselled faces—used for hand axes as well as scrapers and picks. Three-dimensional visualisation was clearly within their skill set, unlike any other creature before them and no non-human creatures since.

Even more tantalising, there is a slim piece of evidence that they created art. A fossilised shell has been found in Java that appears to have deliberate markings carbon-dated to 540,000 years ago. This makes the art too early for Neanderthals or modern humans.[10] As no DNA has been retrieved from *H. erectus* remains, we can't know whether it had one or two copies of the knowledge gene.

Around 250,000 years ago, distinctive oval-shaped and pear-shaped stone tools, technically labelled as 'Acheulean', were being made by *Homo erectus* and *H. heidelbergensis*. The latter is a more recent species that appeared about 600,000 years ago and disappeared some 300,000 years later. Some paleo-anthropologists consider *H. heidelbergensis* to be a subspecies of *H. erectus*. They were visualising three-dimensional shapes to be extracted from lumps of carefully chosen stone. These Lower Palaeolithic tools were fashioned with a symmetry and beauty well beyond what was needed to get the job done. There are even embedded fossil shells that have been deliberately left in the flint, with the tool shaped around them. The best known is the West Tofts hand axe found in Norfolk, England, with its embedded fossil shell of the mollusc *Spondylus spinosus*, as shown in Plate 8.1.[11]

This artistry is not simple when knocking off flakes of flint. It is tantalising to speculate that the knowledge gene was already having an impact. But without DNA, all I can do is speculate.

Despite having DNA evidence that the Denisovans had two copies of the knowledge gene, we have no evidence that they made art. We have no firm evidence that they did anything other than exist. They were first identified in 2010 and there have only been a few bone fragments found so far.

The Denisovans have not yet been granted a full scientific name, but they are members of our genus, *Homo*. They interbred with modern humans, as some of their DNA can be found in contemporary populations. Intriguingly, 40,000-year-old jewellery has been found that some archaeologists have attributed to them. Much older hatch-like markings, dating to 105,000–125,000 years ago, were deliberately made on a bone found in Lingjing, in northern China's Henan Province. Researchers speculate that these may also have been the work of Denisovans. We'll just have to wait for more evidence.[12]

Neanderthals used their knowledge gene

Neanderthals carried two copies of the human knowledge gene. Only a few decades ago, they were considered to be dumb clods living in caves, incapable of speech and certainly incapable of art. However, they moved across the Earth and adapted to varying habitats, providing strong evidence of a knowledge system that can incorporate new information.

In the so-called 'Neanderthal flute', discussed in the previous chapter, we have a hint at their musicality. We have a little bit more than a hint for their art. Palaeolithic archaeologist Rebecca Wragg Sykes is a recognised authority on Neanderthals. She writes that between 135,000 and 50,000 years ago, Neanderthals left traces showing that they, like us, deliberately marked spaces and places. Recently it has been shown that they were almost certainly communicating vocally.[13]

Although no one has found Neanderthal art comparable with the spectacular cave paintings of our species, Wragg Sykes writes about the evidence that their lives were far more complex than was required for mere survival. Evidence of their art includes an etched hashtag in Gorham's cave in Gibraltar. There's a hyena's thighbone with nine parallel incisions on it from France, with evidence of the use of red pigment. There are marks, hand stencils and symbols made with pigments found at multiple sites, with black and red, orange, yellow and even white. A painted eagle's talons and shells add to their gallery. Neanderthals selected particular mineral outcrops that had the most intense pigments, and mixed their pigments, such as adding red to yellow to produce orange.

There is also the fascinating discovery from a German site called Einhornhöhle. A bone from the toe of a giant deer was deeply engraved with a repeating chevron pattern (see Plate 8.2). This engraved giant deer phalanx is at least 51,000 years old. Researchers consider that this object demonstrates that the Neanderthals must have had the conceptual skill to imagine the coherent design as a prerequisite to composing the individual lines. They conclude that it is very likely Neanderthals had an awareness of symbolic meaning and the ability to create symbolic expressions before *Homo sapiens* arrived in Central Europe.[14]

For me, the most astonishing Neanderthal site was first reported in 2018. When this discovery was announced, I received emails from readers of *The Memory Code*, including archaeologists, saying that my 'knowledge spaces' have just been taken back to the Neanderthals. Although I was shocked at the time, I now believe the evidence is compelling.

A cave near Bruniquel, in southern France, has been explored since 1987. Some of the chambers have large, shallow pools, but most surprising of all was the construction in a chamber around 300 metres from the entrance. Initially it was thought to be some kind of dam made out of fallen stalagmites; then it was realised that the masterminds had deliberately snapped off pieces of stalagmite and arranged them. As it dates to between 174,000 and 176,000 years ago, the creators can only have been Neanderthals.

In this remarkable construction, there are over 400 stalagmite sections, weighing in the order of two tonnes (see Plate 8.3). They have been arranged in two rings with an outer diameter of more than six metres. There are two further piles of stalagmites within the rings, and two more piles on the outside. In some places on the rings there are four layers of stalagmite pieces. Some pieces are balanced on top of each other like tiny columns and lintels.

The Neanderthals had used fire to help fracture the stalagmite columns. This is not a habitation site, being way too far into the cave. Known Neanderthal living spaces are found close to the cave entrances. The Neanderthals using this cave must have depended on artificial lighting. A further compelling piece of evidence is that there were no stone tools found in the chamber. Wragg Sykes concludes her description by expressing the difficulties she has understanding how the Bruniquel cave construction could serve any practical purpose.[15]

I have no such difficulty. The Bruniquel Cave construction has all the hallmarks of a memory space, giving it an eminently practical purpose. A huge amount of energy and time was devoted to creating this unique space by Neanderthals, a species who travelled, adapted and flourished for hundreds of thousands of years (far longer than our species has roamed the planet). Slowly we are starting to acknowledge their cognitive abilities. I am convinced that Bruniquel provides the evidence for us to begin recognising a higher level of intellect than we have granted Neanderthals to date. I also predict that other such sites will be discovered, but they're not easy locations to find.

Why did our species create art everywhere they went?

We cannot know the spiritual beliefs that drove early *Homo sapiens* to paint and carve particular images if we cannot hear from them or their descendants about their beliefs. We cannot assume gods or worship or sacrifices, or know if they conceived of an underworld or afterlife. What we do know is that they were working with a brain structure very similar to ours.

We can assume that they took great notice of what was around them, and that they needed to store information if they were to thrive physically and culturally. Those who produced the spectacular rock and cave art were clearly a successful culture.

There is a history of what Australian archaeologist Bruno David refers to as 'Grand Theories' when interpreting rock art.[16] David regrets that these generalisations pay little attention to the particular cultures in the distant past or what we might learn from Indigenous groups today. The Grand Theories started in the 19th century, when all cave art was interpreted as art for art's sake; then it was all about totemism and religious practices and 'hunting magic'; then there was a fashion for binary opposites, such as 'male/female'; and more recently everything has been interpreted as being related to shamans.

The shaman Grand Theory seems to have had a major boost with the work of South African archaeologist David Lewis-Williams, whose books I found interesting on many levels, but unconvincing in their emphasis on shamans and entoptic images, hallucinations and trance states.[17] Paul Bahn, in his influential work *Images of the Ice Age*, also refers to the fashion for 'art for art's sake' and hunting magic theories, and then describes what he calls 'the shamanism impasse'. Acknowledging that shamans were very important figures in parts of Siberia, he argues at length that the transference of that limited ethnography to Palaeolithic art cannot be justified.[18]

Not only do I find Bahn's arguments convincing, but I also feel that shamanistic interpretations diminish the intellectual achievements of contemporary oral cultures. Specifically, Lewis-Williams talks about the San people of South Africa. Research from others makes different interpretations. Andrew Paterson, who studies San art in the Cederberg region of South Africa, writes of two approaches to interpreting the thin red lines on the San rock art:

The currently accepted approach, namely the approach involving the entoptic image, hallucination and trance, advocates that the thin red lines are visual experiences derived from within the eye or brain, not externally as in normal vision. They are part of a set of geometric visual patterns that include dots, zigzags and lines called entoptic phenomena or experiences. All are mental images, manipulated by the San, rather than by the San's intelligence. They are images seen by the San under the influence of psychotropic substances, which when linked to somatic physical responses lead to full hallucinations.

An alternative approach to the above . . . advocates that the thin red lines have been drawn in association with rain, San figures, elephants and eland, deriving from experiences within normal San vision and intelligence. They have been drawn and arranged by the San artist with specific intent, based on recorded San ethnography, regarding the hunter-gatherer's experience of his own reality and perceived interconnected eco-system that includes rain, animals and plants in conjunction with their story-telling, mythology, songs, dances and spirituality.[19]

We should consider ancient human art in terms of a rational intention to store information before we invoke the ubiquitous 'rituals' and 'altered states'. Ross and Davidson write that they 'know of no ethnographic literature that describes the production of art during ritual, nor evidence that ties art produced after altered states of consciousness to the images witnessed during trance'.[20]

I must acknowledge that I am arguing for another 'Grand Theory': that most rock art is created primarily to serve knowledge systems. These obviously include spiritual knowledge. The difference is that my Grand Theory is based on universal observations by oral cultures and offers no specifics on the actual meanings of the symbols.

Humans make art wherever they go

Almost anywhere that humans have set foot, they have left their deliberate mark-making. The vast majority of these marks will not have the material permanence to survive in the archaeological record.

Given the myriad locations on offer, and the range of incredible imagery, I cannot do them justice in one chapter. I have chosen sites that have particular resonance for me. I suggest readers go online and search for images of rock art from any time before literacy and consider how well the argument I offer here holds up. I have no doubt that it will.

Human knowledge is never a static beast. As Iain Davidson has written:

> [I]n societies dependent on oral transmission of knowledge, the production of permanent marks on the environment may allow those marks to be used as mnemonics, but it does not prevent change in the knowledge associated with them. The performance of the ritual may provide the legitimation of the information imparted through the marks, but it does not guarantee the faithful transmission of information from one episode to another. This flexibility is part of the distinctive adaptation of modern humans.[21]

Given that we know humans were genetically encoded to develop this skill set, it is not surprising to see art emerge in increasing sophistication all over the world. The primary concern for all humans is survival. They need to understand their environment, their food sources and what is required to function as a cohesive social group. Without these basic communal information systems, they simply wouldn't have survived, let alone flourished, as they obviously did.

Some of the earliest evidence of artistic endeavour comes from the fascinating South African site of Blombos cave. Perforated shells of a sea snail, *Nassarius kraussianus*, better known as tick shells, found there have been dated to 75,000 years ago. These human-made beads had perforations, indicating that they had been worn as necklaces. The shells had been brought from nineteen kilometres away. There is also worked and engraved bone. More than 8000 pieces of ochre have been found in the cave, some with geometric engravings on them. The earliest have been dated to 100,000 years ago—but the most exciting discovery of all is that the cave was apparently used to process ochre. Two shell containers storing an ochre-rich compound have been found.[22]

Later in the human story artists started to represent what they were seeing. The current earliest known representational artwork in the world is claimed to be a figurative painting of a warty pig (*Sus celebensis*), part of a scene involving other pigs, from the island of Sulawesi, Indonesia, with a minimum age of 45,500 years old.[23]

A continuous art tradition in Australia

As we strive to understand why rock art was created, we in Australia are fortunate to have the most extraordinary asset right here—and Traditional Owners representing many different Nations are willing to help us understand their art.

Current opinion holds that Australian Aboriginal cultures date back at least 65,000 years, in a continuous, but not unchanging, cultural expression. We know that stories told today make specific reference to events that happened well over 10,000 years ago. That is in the order of a few hundred human generations, through which particular items of knowledge were valued so much that they were reliably handed on. I find that simply extraordinary.

We know that ochre was being used over 40,000 years ago, as it has been found in the burial of 'Mungo Man' at Lake Mungo, in south-western New South Wales (see Plate 8.4). Damian worked there with Traditional Owners as an archaeologist, and described his experience for me:

> During a lot of archaeological surveys you find evidence of the animals who are now extinct in the area such as hairy-nosed wombats, swamp rats, bettongs and *Genyornis*. These species simply couldn't survive the huge change in climate from a ten-metre-deep lake to what is now desert. But the Aboriginal people remained on the site throughout these enormous climatic changes. How did they do that? Because they were way more adaptable than the other species. They knew how to use resources and adapt to new ways of doing things.
>
> Evidence from otoliths (once part of a fish's inner ear) indicate the growth changes over time, which charted some of the decline in water levels as fish got smaller and had different growth patterns. Using the evidence from camp-sites, people changed diet completely from a lot of aquatic animals and fish and tortoises to dry country species such as reptiles and kangaroos.

Unfortunately, given the nature of the sandy environment, art is unlikely to have survived. There are no caves or rocky outcrops where any sort of permanent art could survive. There is evidence of art from the iconic Aboriginal skeletons which include some with ochre. Ochre came to the area from further afield, so clearly it was being traded.

There is evidence of the use of ochre dating back over 30,000 years in the Central Desert area of Australia. The rock art of Central Australia shows no identifiable stylistic breaks; there is a continuity of essential motifs being used, which demonstrates that there is an enduring graphic tradition across the millennia.[24]

Wherever there was a suitable rock surface—be it a shelter, boulder or rock platform—Australia's First Nations people marked it with art. Consequently, Australia has a rock art inventory as large as that of anywhere in the world. It consists of both pictographs (markings painted onto the rock) and petroglyphs (markings carved into the rock).

The images are abstract and figurative, with many dating back tens of thousands of years. Often the new marks build on earlier marks, signifying a continuous tradition. We know from contemporary Traditional Owners that this art reflects their connection to Country, their spirituality and ancestral beings, alongside history, law, genealogical relationships, intertribal relationships and their relationships with plants and animals.[25]

A large proportion of the designs are geometric motifs, such as circles, arcs, lines, spirals, dots and animal tracks. Abstract symbols enable multiple interpretations even within a single image. The interpretation can depend on both the context of the storytelling and the level of initiation of the artist and of those permitted to view the image.[26] Australian restricted and sacred art is almost exclusively abstract designs, including animal tracks, while the secular art in domestic situations tends to be more figurative.[27]

Using abstract art to represent knowledge to be restricted to those appropriately initiated is found in First Nations cultures around the world.[28] Not surprisingly, then, there is a correlation between the more abstract symbols in rock art and physically restrictive locations globally.[29]

I first became aware of one of the most stunning examples of Australian Indigenous art through the 2013 television series *First Footprints*.[30] There I saw the magnificence that is Nawarla Gabarnmang, one of thousands of rock art sites in Jawoyn Country, in Arnhem Land in Australia's Northern Territory. Aboriginal occupation dates to at least 45,000 years ago.

So much is revealed as the archaeologists discuss the site with Traditional Owner Margaret Katherine. Nawarla Gabarnmang was fashioned by people systematically removing large sections of bedrock.[31] Above the twenty sparsely distributed pillars is a vast flat ceiling, which served as a canvas for the most extraordinary artwork, created layer by layer over the millennia of use (see Plate 8.5). For Jawoyn people today, the site and art are permeated with powerful spirit-beings, some of whom have metamorphosed into the rock and can be seen in the paintings while also remaining alive, still wandering across the landscape as keepers of Country.[32]

A staggering 1391 paintings and stencils have been identified on the ceiling, with hundreds more on the pillars.

So far, the earliest art has been dated to nearly 30,000 years ago. Some panels show up to 43 layers of artwork. The latest style includes the so-called 'X-ray art', which was still being practised in Arnhem Land when the first Europeans arrived. There is even a horse in one panel, a clear indication of adaptation to new knowledge, the first horse having arrived in the area only in 1845.

Nawarla Gabarnmang offers an insight into millennia of continuous and dynamic culture. As the Traditional Owners explain, this artwork tells the stories from which they learn and adapt and, through that knowledge, survive. Archaeologist Bruno David describes Nawarla Gabarnmang as a 'grand, architectural monument that was also a complex, three-dimensional artistic canvas'.[33]

The artists at Nawarla Gabarnmang created a performance space unrivalled in the world. And the Jawoyn are still performing, dancing, creating music, learning and teaching.

On another side of the continent, a vast amount of art was being produced at Burrup Peninsula, in north-western Western Australia. Recent research dates the ever-changing art styles from over 50,000 years ago until less than 3000 years ago.[34] Again, there has been a continuous tradition from then until now, which can help us understand how the motifs were used.

Burrup Peninsula, previously known as Dampier Island, is also known by its Aboriginal name, Murujuga. It is a stretch of land about 30 kilometres long and six kilometres wide, which, before industrial development, was separated from the Western Australian mainland by mangroves and tidal mudflats. The hardness of the rock preserved the changing artistic patterns over tens of millennia. In 2011, archaeologist Ken Mulvaney prepared a detailed survey of the area, working closely with Traditional Owners.[35] There are as many as 6000 sites, with between 500,000 and 1 million petroglyphs, probably the greatest concentration of petroglyphs in the world. Images of humans performing in ceremonial dress link to contemporary practice.

There are cupules, geometric designs, tracks and figurative images. There are turtles, fish, macropods, birds, reptiles, thylacines, echidnas, crustaceans and stingrays. Mulvaney described the location as originally a vast open plain with hills and ranges. It became a series of coastal islands close to the shore. Consequently, as the art sequence progresses, marine species become more dominant. For example, the birds shift from predominantly terrestrial to aquatic species.

In this later phase, after the rises in sea level about 7000 years ago, two distinctive motifs become common. These are shown in Figure 8.1; how would you interpret them?

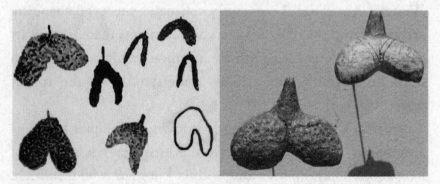

FIGURE 8.1 Left: Symbols from rock art at Burrup Peninsula identified by Traditional Owners as stingray and shark livers and as the tail of dugong or whale. Right: Portion of display labelled as 'Female figurines from Dolní Věstonice'. PHOTO: ZDE (CREATIVE COMMONS LICENCE CC BY-SA 4.0).

Local Aboriginal informants identified the first shapes as stingray or shark livers. The shapes in the centre have been identified as whale or dugong tails.[36] On the right of Figure 8.1 are objects identified as breasts from Dolní Věstonice, an Upper Palaeolithic archaeological site in the Czech Republic.[37] But are they breasts? We can't know for sure what abstract shapes meant to the people who created them so long ago. It is likely that without the interpretation of the knowledge keepers, the breast-like shapes found on the Burrup Peninsula would also be labelled as breasts.

Palaeolithic artists still manage to astound us

In over 400 caves, Upper Palaeolithic people in Europe created non-figurative signs from at least 42,000 years ago and figurative images from at least 37,000 years ago. Art survives in caves and shelters far better than it does in outside galleries. The decorated caves and shelters we know about in Europe represent only a small proportion of what would have been created over the 30,000-year span of the Upper Palaeolithic.[38]

Canadian author and paleoanthropologist Genevieve von Petzinger explored almost 400 sites across Europe, identifying the 32 most frequent signs in Ice Age cave art. These same signs were used for 30,000 years, although she recognises that their meanings may not have remained constant.[39]

Also analysing a huge database of signs, other researchers have hypothesised that three of the most frequently occurring signs—the line, dot and Y-shape—created a system of associating animals with calendar information to record and convey the seasonal behaviour of the species they hunted. The line and dot refer to lunar months and the Y to the time of birth.[40] Even though we can't know if this interpretation is accurate, I am delighted to see researchers hypothesising that these artists, who were behaviourally and intellectually as capable as anyone alive today, were storing information. The time has come for a rational, pragmatic Grand Theory for cave art.

Much of the art was deep in cave systems, often hundreds of metres from any natural light source. The artists, and any viewers, used Palaeolithic lamps fuelled by animal fat.[41] Decorated caves were not places in which people lived; their dark, slippery interiors, with lots of hard sharp bits sticking out, made for an unappealing home site. During the Palaeolithic, most people lived in the open air.[42]

You will not be surprised that I'm going to argue that these caves were restricted knowledge sites. The fact that there are occasional handprints from children, and even their foot-prints, does not imply that the information maintained and conveyed there was not restricted. Those children may well have been the offspring of the artists, accompanying them as the knowledge was being painted on the walls. Or maybe they were the children of a knowledge elite being prepared for their future role. Performance spaces for small groups at a distance from normal living spaces are typical of Indigenous knowledge systems.

Not all Palaeolithic art is found on cave walls. It is estimated that there are well over 10,000 items identified as Palaeolithic portable art in Western Europe alone. This estimate, dating back to 1980, is certain to represent only a fraction of the portable items created. Although many European Palaeolithic sites have no decorated objects, some have thousands. In the

Spanish cave of Parpalló, an astonishing 5034 engraved and painted stones were found.[43]

The French cave of La Marche has contributed more than 3000 engraved stones to the statistics. Unusually, these include numerous representations of people.[44] Although many were portable, some of the blocks of stone weighed up to 50 kilograms. Some had been brought from about 30 kilometres away.[45] This was a purposeful operation. Many contemporary Indigenous cultures use collections of objects to help them recall sets of information and sequences in ceremony.[46] It is perfectly logical to conclude that Palaeolithic people were doing the same.

Palaeolithic artists were exceptional in utilising their visuo-spatial skills, so closely linked to the knowledge gene. They would incorporate naturally occurring shapes in the walls. They would paint on all sides of rocks, in crevices and around corners, even encircling cylindrical bones with wonderfully realistic representations of animals.

Sometime between 18,000 and 15,000 years ago, one such artist carved a reindeer horn with images of deer and leaping salmon in the French cave at Lortet. They used a technique known as *champlevé*, which involves scraping away the bone around the figure so the design stands out, as with a cameo. It is astonishing to realise that he or she managed to do this and retain perfect proportions in the animals even though the whole figure could never be viewed at once.[47] Drawings of the carving and unrolled versions were published by the British zoologist Sir Edwin Ray Lankester in 1920 (see Figure 8.2).[48] That artist used their innate visuospatial skills with flair.

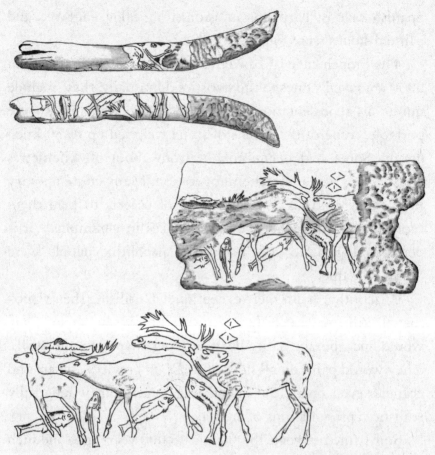

FIGURE 8.2 Drawings of the carving and unrolled versions of the Lortet reindeer horn were published by British zoologist Sir Edwin Ray Lankester in 1920 in *Secrets of Earth and Sea*. IMAGE: LYNNE KELLY.

Figurines are everywhere

There is only one species with the visuospatial skills to create figurines from materials that bear no resemblance to those shapes. Humans do that all over the world, and have done so for millennia. In contemporary First Nations cultures, figurines often represent the mythological beings who act out the stories.

As mentioned in Chapter 4, I have long been intrigued by the way the Pueblo of the south-west United States have represented their mythological beings as *kachinas*. These characters are created as figurines, each with a distinctive face marking and costume. The masks and costumes are also used by dancers in performances, while representations are found on pottery and in rock art. Hundreds of *kachinas* form a complex, ordered set of characters, appearing in ceremonies in a strictly controlled sequence over the year. Kokopelli, the hump-backed flute player, can be seen in rock art dating back a thousand years (see Figure 8.3), through to contemporary art. He is, in part, a fertility figure.[49]

FIGURE 8.3 Petroglyph of the flute player Kokopelli, in the Mortendad cave, near Los Alamos, New Mexico, USA. PHOTO: LARRY LAMSA (CREATIVE COMMONS LICENCE CC BY-SA 2.0).

On my shelf, alongside my collection of contemporary *kachinas*, is a clay figurine of a woman with many children. She is a common image among Pueblo cultures. Without the associated oral tradition, I dare say she would be identified as a fertility figure. She is not. She is the Pueblo storyteller, representing the importance of stories in conveying knowledge.

Nude female African Luba sculptures are interpreted as fertility figures, despite evidence from the artists themselves that they are not about fertility; nor are they 'ancestor' figures, the other common assumption. A Luba female figure holding or gesturing to her breasts (see Plate 8.6) represents the belief that only women are deemed strong enough to keep the profound secrets of royalty. Within their breasts, they protect the royal prohibitions on which sacred kingship depends.[50]

In Africa, the Senufo of southern Mali use a set of 58 figurines to represent animals, people or symbols of activities. These are shown to novices in a prescribed order. Elders use the figurines as a set of meaningful symbols, which French anthropologist Claude Lévi-Strauss describes as a 'canvas of instruction imparted to them'.[51]

Female figurines across European Ice Age sites are usually referred to as 'Venus figures', such as the Venus of Willendorf and Venus of Laussel (see Plate 8.7). Venus was the Roman goddess of love and beauty. Many of these figurines have drooping breasts and sagging stomachs but were still labelled as Venuses by archaeologists who were likely to be male. Many appear to be more like the drooping naked form of older women. Matriarchs are also important in all societies.

The upturned horn held by the Venus of Laussel is sometimes interpreted as a musical instrument, while other researchers consider the thirteen lines carved onto it to represent lunar and menstrual cycles. Of course, it could be both. Maybe, with my own biases, I should interpret these older women as leaders in matriarchal societies across Europe. We simply don't know, and we can't know. Their meaning is beyond our grasp.

There are male figurines as well. Considered the oldest known figurine, the *Löwenmensch* is a prehistoric ivory sculpture discovered in the German cave of Hohlenstein-Stadel and dating to somewhere between 40,000 and 35,000 years ago. *Löwenmensch* is German for 'lion-man': the figure has a lion's head on a human figure (see Plate 8.8). Mythological creatures with combined human and animal characteristics are common characters in oral cultures, often acting as metaphors for human behaviours. To see such a figure from so very long ago is not surprising, but further interpretation is impossible.

Europe's extraordinary Ice Age cave art

Art in Ice Age caves seems to be produced in panels.[52] Photos of the most gorgeous horses and bisons show the incredible skill of some of the ancient cave artists. These images are stunningly beautiful. But there's some pretty ordinary stuff in there, too. When you observe the entire panel, the art starts to look a little different.

We have art galleries in Australia that have been used continuously since well before the famous French caves of Lascaux and Chauvet and Spain's Altamira. I feel justified in assuming that those who painted the incredible cave art there were just as intellectually advanced as their Australian contemporaries.

The oldest of the famous threesome is Chauvet, discovered in 1994 in south-eastern France and immediately protected before tourists could wreak havoc on its precious walls. Grotte Chauvet-Pont d'Arc, as it is formally known, offers us the best-preserved cave art in Europe.

I cannot even attempt to describe the glorious offerings in Chauvet. I highly recommend you head online to seek out the images. There is also a superb 2010 film, *Cave of Forgotten Dreams*, which offers glorious insight and some idiosyncratic interpretation by Werner Herzog.[53]

Instead of looking only at selected images of superb artistry, try to find images of entire panels. Take the famous horse panel (see Plate 8.9), and ask yourself: is this art for art's sake? Is this just idle doodling? Or is this an educational canvas? Look in particular at the four horse heads. Equestrian experts will tell you how much can be understood from the position of a horse's ears. The artwork clearly illustrates the different positions of the ears and probably other behavioural observations. The fact that it demonstrates masterful artistry is a bonus.

Similarly, a great deal can be told from the condition of a horse's coat over the seasons, another detail that seems to be indicated here. The more you study this panel, the more you can believe that it was used to record and explain animal behaviour. Both male and female lions feature in massive panels at Chauvet. The lions' heads are depicted in different positions, an indicator of the lion's awareness of humans. A rhinoceros head is shown in movement. Clearly these artists were capable of representing the entire animal, as they did elsewhere, but it was critical for them to understand the various signals of the animal's behaviour. The heads give the information that is needed.

Chauvet is a complex and extensive cave system (see Figure 8.4). We need to think of the physicality and placement of the art when asking if this is a knowledge space.

Beyond the artwork, Chauvet includes many features that indicate its use as a knowledge space. There are structures created by the occupants, including piles and alignments of rocks. Most revealing is the skull of a cave bear (*Ursus spelaeus*), which has been intentionally placed on a large stone block. It has been dated to just over 30,000 years ago from charcoal

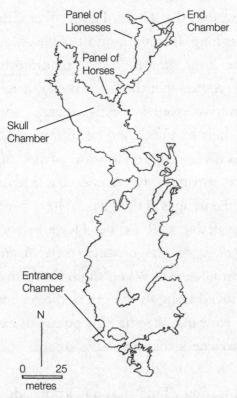

FIGURE 8.4 Map of the Chauvet cave. Adapted from J. Delannoy, B. David, J. Geneste, M. Katherine, B. Barker, R.L. Whear & R.B. Gunn, 'The social construction of caves and rockshelters: Chauvet cave (France) and Nawarla Gabarnmang (Australia)', *Antiquity*, vol. 87, no. 335, 2013, p. 14.

found on its outer surface. The skull is the centrepiece of a complex configuration involving dozens of cave bear skulls and associated decorated cave walls.[54]

Interesting results come when people from different disciplines offer expert insight into the interpretation of the cave art at Chauvet. Marc Azéma is a Palaeolithic researcher and filmmaker who has spent more than twenty years exploring the representation of animal movement in cave art. He argues that the primary function of these cave paintings was narrative. For Azéma, the cave was a theatre.[55]

Azéma describes the 'Grand Panneau' at Chauvet, a frieze over ten metres long. It contains most of the species in the cave, including cave lions, horses, bison, mammoths and woolly rhinoceros. He argues that the frieze probably tells the hunting story through two events. Firstly, several lions are shown stalking with their ears back and heads lowered, and secondly, a pride of lions is seen lunging towards a troop of fleeing bison.

Even more intriguingly, Azéma has identified signs of movement in the images as they would have been viewed with the lighting available at the time. He talks about the superposition of successive images, such as with an eight-legged bison drawn in Chauvet. When the light from a grease lamp or a torch is moved along the length of the rock wall, the bison appears to be running. Azéma also points to examples, often on portable artworks, of successive images showing animal movement.

This is education. This is about knowledge. This is not pretty pictures for the sake of aesthetics.

The knowledge gene links musical and visuospatial skills, so it is not surprising that many authors have linked the position of art in caves with acoustic responses. One such study claims that rock art was placed in caves and canyons particularly selected for their intense echoes. The researchers argue that 'more than 90 per cent of European cave art depicts thundering stampedes of ungulates, located in portions of caves where a single clap results in thunderous reverberation sounding like hoofbeats'.[56] With rhythmic clapping, vocalisation and possibly the use of musical instruments, the impact must have been extraordinary.

At Spain's Altamira cave, there are further spectacular examples of cave art. Again, they fit the pattern of knowledge symbols in a performance space. For example, there is a colourful ceiling showing animal behaviour (see Plate 8.10). A detailed survey of the ceiling can be seen in Figure 8.5.

These caves were painted by intelligent people who knew how to learn and how to teach.

Reluctantly, I have to limit the scope of this chapter, much as I would adore talking about every Ice Age European cave

FIGURE 8.5 Survey of the polychrome ceiling of the Altamira cave, published by M. Sanz de Sautuola in 1880 (after Cartailhac, 1902).

with stunning art. Any search for the images risks leaving you speechless. The stories of the discoveries and the debates around their art are also fascinating, but they are reported extensively elsewhere.[57] My focus is on knowledge systems. But I must mention one more specific example, this time from France's Lascaux cave.

More than 600 paintings and nearly 1500 engravings are found at Lascaux, dating to at least 15,000 years ago (although that is still being debated). The fact that pragmatic knowledge must have been stored by the people who created this extraordinary site can be seen in the analysis of their pigments. There were plenty of stored pigments ready for use—black, yellow, red and white. Some came from close to the cave but there was also manganese dioxide, which had been brought from five kilometres away. But they didn't use the pigments straight from the ground. They had recipes for creating the colours they wanted. The colour of ochre is modified by heat, and Palaeolithic people clearly knew this.

One pigment contains calcium phosphate, which the artists must have created by heating animal bone to 400 degrees Celsius. The result was then mixed with calcite and heated again to 1000 degrees, which transformed it into white tetra-calcite phosphate. Other complex mixes have been found at Lascaux and elsewhere.[58] It was a much more glamorous location than our contemporary chemistry labs.

Before I leave Europe, I must acknowledge that other researchers also claim that the continent's rock art is all about narrative and knowledge. For example, archaeologist Knut Helskog described rock panels in northern Europe as demonstrating a story covering the entire annual cycle, including things like seasonal aggregations of animals, with invaluable seasonal indicators such as coat condition and the shape of horns.[59] It is lovely to be part of a growing group asking that pragmatic interpretations be considered.

To the Americas for spectacularly long art galleries

Right across the Americas, there is art that follows similar patterns. I can only include a few examples, but in my view they indicate why we should consider rock art in panels, rather than as the individual images we might extract. In Chapter 11, I will propose that we bring exactly this type of knowledge technology into our classrooms.

In the Amazon rainforest of Colombia, there is a rock wall canvas running for almost thirteen kilometres. Mastodons, giant sloths and other extinct beasts roam across the surface at the Serranía de La Lindosa archaeological site. There are geometric designs and handprints amid the deer, alligators, bats, tapirs, turtles, serpents, camelids, horses and three-toed hoofed mammals with trunks.

Plate 8.11 shows just a small portion of the extraordinary stretch of red-ochre images, the earliest of which, researchers estimate, were painted around 12,600 years ago. Additions date right up to European arrival.[60] This is an Ice Age blackboard.

If you think a thirteen-kilometre-long art gallery is incredible, then consider Nine Mile Canyon in eastern Utah, United States. The canyon is actually a lot longer than nine miles (14.5 kilometres). The Sandhill Crane Site at the mouth of Currant Canyon, as shown in Plate 8.12, is a tiny portion of the entire site. The rock art extends for a staggering 72 kilometres, and is promoted as 'The World's Longest Art Gallery'.

In some places, the ancient artists etched their images from incredibly narrow, high cliff ledges where they could easily slip to a certain death. Archaeologists estimate there are 75,000 to 100,000 images in the canyon.[61]

On a different surface, panels of art still offer the same mnemonic technology. Carved deeply into boulders of a partially collapsed tufa mound dating to over 10,000 years ago, the petroglyphs at Winnemucca Lake subbasin, in Nevada, are North America's oldest (see Plate 8.13).[62]

While we're in the United States, I must talk about a site that had a profound impact on me. I've come across many things I have initially struggled to explain but for which, with time, I could always find a pragmatic, rational explanation. Except this time.

In 2009, I was in New Mexico to study contemporary Pueblo cultures and Ancestral Puebloan archaeology. On the outskirts of Albuquerque is the Petroglyph National Monument. On the hilly site, Ancestral Pueblo carved most of the approximately 24,000 images into the soft volcanic rock. On the drive to one of the walks, the interpretive officer gave me two warnings. Firstly, watch out for rattlesnakes. And secondly, some of the petroglyphs will only appear if they want you to see them. To her surprise, she had experienced this effect, as predicted by her First Nations advisers.

A sign indicated the presence of a spiral on a rock. There was no spiral. I looked from every angle. It was not there. I photographed the unadorned rock. Half an hour later, I returned and the spiral was perfectly distinct. I photographed it again. In both sets of photographs, the spiral was clearly present. I cannot explain what happened. But then again, as a scientist, I have to acknowledge there is a lot that we can't explain. What can be explained from the very detailed report about the monument is that the site works exactly like other knowledge spaces.[63]

Further north on the American continent, there are still more examples of vast panels of art. For example, Canada's Petroglyphs Provincial Park, in Ontario, contains around 1200 carvings, incised deeply into the soft walls up to 1000 years ago. There are humans, animals and a figure taken to represent the sun. Tellingly, the First Nations people of Ontario refer to the carvings as *Kinomagewapkong*, which they translate as 'the rocks that teach'.[64]

And that's what I believe they all do. The petroglyphs, the carvings, the paintings—they all teach.

Rock art galleries are a universal pattern

There is a site in Africa that I found as astounding as any of those mentioned to date.

The UNESCO World Heritage site of Tassili n'Ajjer, in Algeria, is one of the most spectacular in the world.[65] There are more than 15,000 drawings and engravings, painted from about 8000 years ago until the present era, and recording climatic changes, animal migrations and human activities. The site contains a carving that I find extraordinarily beautiful, the sleeping antelope shown in Plate 8.14.

Although I have been making the case that prehistoric rock art is primarily about knowledge, that is not to deny the aesthetic sensibilities of Palaeolithic people. I would be hugely surprised to find out, which of course I never can, that they did not look at the carving of a sleeping antelope and appreciate its beauty. And is there any chance that they did not recognise the extraordinary skill of the artist?

But it must be remembered that this stunning creature is part of a massive rock art gallery. The drawing in Figure 8.6 represents just a tiny sample of what is found within the Iheren rock painting shelter in the Tassili n'Ajjer mountains in south-east Algeria.

The same patterns can be seen right across Asia, too. To explore glorious rock art images for Asia, and the world as a whole, I cannot recommend the Bradshaw Foundation highly enough. It is a non-profit organisation focusing on archaeology,

FIGURE 8.6 Part of the massive panel at Tassili-2, and a detail showing the giraffes. IMAGE: LYNNE KELLY.[66]

anthropology and genetic research, and has a superb website offering numerous images and excellent commentary.[67] There are particularly strong sections on India and China.

The art of India constantly changed form, from ancient art panels in caves and on rock faces, through to long, narrative, symbolic representations of knowledge that are still in use today. Particularly illuminating are the Patuas, performers who live in West Bengal. As artists, they produce narrative scrolls formed from a series of framed images. These tell stories ranging from ancient Hindu myths to current events. For each story, the Patuas compose a song, which they perform as they gradually unroll the scroll and reveal the narrative. As with Indigenous cultures everywhere, the performers combine entertainment with education, music with art and storytelling.[68]

Exactly the same format is found with the traditional performing arts of China. I'll return to that story in Chapter 10, as I explore what happened when literacy usurped the Chinese performers' role.

Art for knowledge's sake

As we'll see in Chapter 10, it is only recently—well into the days when literacy had become widespread—that the role of art significantly changed from being primarily for knowledge's sake to something that is primarily aesthetic.

When viewing the prehistory of art, meaning and intent is lost somewhere in the abstract and jumbled motifs, the images superimposed over each other, the vast rock art panels and the connections to music and performance spaces. Despite this, I don't believe there can be any conclusion other than that these art galleries actually embody knowledge.

We can look to the past and recognise an untapped potential for our future.

The cultures that created the incredible art spaces have a great deal in common with each other, despite being separated by vast distances of both time and place. But they are all working with the same human brain. Creating art is a biologically driven instinct, innate, universal, uniquely human and incredibly ancient. Art is inspired by genetically driven urges, and what underpins all of this is the knowledge gene.

Intriguingly, a 2015 study indicated that engaging in art and crafts in old age can lower the risk of developing mild cognitive impairment and the early signs of dementia by 73 per cent.[69] Hospitals in the United Kingdom are applying this research, using art as part of their preventative actions with older patients.[70] But surely we should all be taking advantage of these clues to the way our brain naturally functions.

Producing knowledge art requires cognitive effort, accurate observation, interpretation and the creation of an image in two or three dimensions. Even if the art wouldn't sell in a commercial gallery—even if it is merely stick figures and abstract symbols—the process can be invaluable. It is the process that stirs the brain to lay down neural pathways. That is available to all of us.

These same cognitive processes can be seen in the contemporary art of Indigenous cultures, when the art is produced primarily for knowledge's sake. Creating sculptures and representations of narrative demands cognitive effort, focus and the use of the visuospatial skills so clearly enhanced by the knowledge gene NF1. The myriad art forms in our long prehistory can offer models for us to use in contemporary life. With no negative side effects, and so much to gain, the evidence is clear that sidelining art from our ways of knowing is a waste of our potential.

I'll return to that theme in Chapter 11, but first I want to explore the evidence that our spatial skills, so wonderfully championed by our knowledge gene, served communities globally when they grew and settled. So valuable was the use of space to store knowledge that communities invested unimaginable effort to construct monuments for that very purpose.

CHAPTER 9

The monumental story of knowledge spaces

The art, music and decorated spaces described in the previous chapters served humans well over our long prehistory. Some communities grew larger and settled, and during that transition they built enigmatic monuments. People built monuments for a purpose, a purpose so significant that the communities were willing to invest millions of hours into constructing them. What purpose could be more important than preserving essential practical and cultural knowledge? That's what the monuments were all about.

The only species on Earth who built monuments were humans, because they were the only species who were born with the massive advantage of their knowledge gene, NF1. They didn't just rely on knowledge stored through their artistic or musical or performance skills, nor on their ability to conceive ideas in three-dimensional space. They created monuments that enabled optimal use of all of these in combination.

We don't know the beliefs of these people, or their laws and regulations, expectations for life and individual ambitions. But what we do know is that they must have stored vast amounts of pragmatic information in their oral tradition. We can see how they enhanced their memories using every possible memory device in the designs of their knowledge spaces. That is why there is an underlying pattern in the monuments built by oral cultures all over the world.

It is usual to study these monuments in isolation. The British Neolithic with its incredible Stonehenge, Rapa Nui (formerly Easter Island) with its amazing *moai*, or Peru with those phenomenal glyphs in the Nazca desert. But studying them in isolation severely limits the insight we can gain by acknowledging that they were all built by humans using the same brain structures, with the same genetics and with the same physical, cultural and spiritual needs. By looking at the monuments as a group through the lens of knowledge systems, we can see that wonderfully individual implementations served the same purpose. All human societies need knowledge spaces.

What happened before the monuments?

The monuments I'll be talking about in this chapter are all from the last 12,000 years. The previous two chapters show that humans were using music and art and their visuospatial abilities as far back as 40,000 years ago. As humans began to congregate in larger groups, and eventually settle in permanent communities, things started to change.

Some societies were no longer travelling the landscape,

their vast memory palaces. The transition from mobile communities to those staying still to work their farms did not happen overnight. Monuments were built and constantly changed during that transition, which took thousands of years.

We have tens of thousands of years of continuous Australian Aboriginal culture, with robust evidence of the use of material knowledge technologies for well over 10,000 years, long before any of the monument builders had been born. We know how they were using spaces and art and music as part of their knowledge system because they've told us so.

In south-eastern Alaska, we know that the stories of the Tlingit, Haida and Tsimshian peoples date back well over 10,000 years. We know that they used their art spaces and incredible totem poles to store information because they've told us so.

Those in Australia and those in Alaska were so far apart, separated by great stretches of water, that there is no possibility they were in contact. Yet they use the same memory technologies. The common factor is human biology, human cognition and human intellect. Should we assume that the other cultures on the planet, those who built the monuments, were somehow less engaged intellectually?

It is worth recalling that Stonehenge was built only 5000 years ago. If we make one simple assumption, many monumental mysteries become far less enigmatic. That assumption is that they used the knowledge methods for which they were genetically encoded. Humans seek patterns and store all that they learn.

I have written before about famous monumental spaces as far back as the European Neolithic and American Archaic.[1]

I chose to research forward, to see what happened as literacy aggressively obliterated most traces of orality. That turned out to be a fascinating story, too, one I will tell in the next chapter. Thinking about the knowledge gene forced me to take the journey as far back in time as archaeology would let me go, as I have in the previous two chapters. Now is the time to explore another global pattern.

In *Knowledge and Power in Prehistoric Societies* (2015), I assessed whether a monumental space could be considered a knowledge space by asking for at least eight of ten indicators to be met.

The ten indicators of a knowledge space

The first indicator is that, in the earliest stage of monuments, we should see evidence of a stratified society, yet no sign of individual wealth. That is because power in small-scale oral cultures was held by those who controlled knowledge, who formed a knowledge elite. It is only when you start to get much larger communities that grave goods appear in individual burials, a sign of individual wealth.

Similarly, in the early stages of the monuments, there were no weapons buried with individuals. It is only with larger societies that power was transferred to those who used wealth and violence to maintain their dominance.

Oral cultures are often referred to as egalitarian, sharing all they have. That is only in reference to material goods. There is no society that shares knowledge equally among all members of the community. If there was, that information would soon be corrupted.

For that reason, the second indicator is the presence of both public and restricted ceremonial sites. As described in

Chapter 3, small and large sites are needed for performances of both public and restricted knowledge. Performance spaces may take the form of platforms, mounds, enclosed areas, plazas and flat-bottomed ditches, or any other space that has clearly been designated as important but that serves no purpose for residences or manufacture.

Communities of people do not spend thousands of hours creating a monument just for the fun of it. The third criteria is the expenditure of vast amounts of energy for no apparent reason (unless you are already thinking that performing knowledge is a really good reason).

If a knowledge space is to be used as a sophisticated memory palace, then there has to be some evidence of a prescribed order. There must be some set sequence to the monumental structures, such as posts, stones or mounds. This fourth criteria is easy to falsify, as any decent hypothesis must be. If you can find a Neolithic monument in Britain or Western Europe or an Archaic monument in the Americas with a random layout of standing stones, timbers or mounds, then the hypothesis has a major failing. I have yet to locate one, despite having searched thousands of sites. A circle is the obvious arrangement to represent the annual cycle of rituals, so it is not surprising that we find stone and timber circles all over the world in the monuments built in this transitional stage.

The fifth criteria is the presence of what are often called something like 'enigmatic decorated objects' or 'ritual objects'. From Chapter 4 you will be well aware that most, if not all, oral cultures use such objects as memory devices.

Knowledge is traded. Songs, dances and entire ceremonies are traded, carrying the knowledge they represent. At the monuments there are stones and timbers, often brought from far away. There are often valuable objects along with the food and other necessities required by the people building the monument or engaging in the ceremonies. If nothing of material value is leaving the space, then we have an imbalance of trade. The balance is restored if you include knowledge in the equation. Criterion 6 looks for this apparent imbalance in trade.

Criterion 7 looks for physical signs for pragmatic information being recorded—in particular, astronomical alignments. All cultures use astronomy to maintain calendars for seasonal, agricultural and ceremonial purposes. In oral cultures the world over, individuals with astronomical knowledge are held in high esteem and granted significant power. A single stick in the ground will allow you to track the sun; you don't need an entire monument. So although it's a valuable indicator of a knowledge space, astronomical alignments are not a justification on their own.

As shown in Chapter 4, the landscape provides a sophisticated memory palace for cultures whose members move around their territory and know it well. Criterion 8 requires that monuments reference that broader landscape. As a society becomes more localised, settling into an agricultural lifestyle, they cannot afford to lose the information stored in their sacred places around the broad landscape. The solution is to replicate the landscape journeys locally.

To take full advantage of the memory systems available to our human brains, we need music. The best indicator of music is some sign of acoustic enhancement, as is required to satisfy the ninth criterion. The monument builders wanted their music to resonate loud and clear, and have a dramatic impact. Testing should include that universal instrument, the human voice.

The final indicator, in tenth position, is one that you might think should be higher up. Rock art is a memory knowledge source of huge importance, but many landscapes do not have caves or convenient rock surfaces. Petroglyphs may survive, but outside surfaces are likely to have been painted. We won't see those painted images, which will have eroded away long ago. This is the hardest indicator to rely on.

Any interpretation of an archaeological site that ignores the intellectual world of those who created and lived in it is an impoverished interpretation.

It is critical to note that human cultures were not on a neat pathway from hunter-gatherer-fishers through to modern Western civilisations. This assumed trajectory is merely an artefact of archaeological interpretation during the last century by some Western archaeologists. Most do not accept that linearity today. Another falsehood often presented in older archaeology texts is that monumental architecture was only possible among agricultural societies. The sites of Poverty Point and Göbekli Tepe, discussed below, quickly dispel this myth.

Posts, poles and standing stones

To create a memory palace, each location within it needs to be marked. When the cultures were moving around their broad homelands, they used natural features, sometimes enhanced with cairns or markings on the rocks and timbers. In order to localise these memory locations, the obvious thing to do is arrange stones and timbers in sequence. Each stone needs to be unique, as they are naturally. The tall timbers can be carved and painted, just as the totem poles are for the Pacific Northwest Coast cultures today.

Tlingit Elder David Kanosh's words might inspire you to imagine how arrangements of posts or non-residential buildings might have been used long ago:

> Totem poles dedicated to origin stories/creation cycles remain purely for that story and I encode no clans nor genealogies.
>
> Totem poles which would designate the territory of a clan, I encode strictly the clan and clan houses for that alone. I do not encode genealogies. Clan house totem poles which tell the origin of that very house, I will thus encode that story and a list of clan house leaders.

I am not the first to suggest that totem poles can be seen as a model for wooden posts in the archaeological record. The oldest known wooden sculpture in the world was made during the Mesolithic period, nearly 12,000 years ago. Currently in the Sverdlovsk Regional Museum of Local Lore in Yekaterinburg, Russia, the Shigir Idol (see Plate 9.1) was found in a peat bog in 1890, a rare example of such long-term preservation of a wooden object. The full height of the sculpture was originally 5.3 metres.

The Shigir Idol is probably not as unique as it might seem. There is plenty of evidence of large, worked logs from the time. German archaeologist Thomas Terberger and his team suggest that the totem poles of the Haida in Canada might give us an idea of the use and role of such sculptures. They note that the simple lines and zigzag patterns on the sculpture are common designs in Late Palaeolithic and Early Mesolithic decoration. There are similar monumental anthropomorphic figures at Göbekli Tepe, in eastern Anatolia. Over such a great distance, there is no suggestion of contact between the two cultures.[2]

Around the world in mnemonic monuments

When viewed through a knowledge gene lens, 'enigmatic' monuments become so much less mysterious. The monuments were theatres where knowledge was performed with gusto and flair. Large theatres and small intimate theatres, but they were all theatres. This pattern can be seen the world over.

Göbekli Tepe is considered one of the oldest examples of monumental architecture in the world (see Plate 9.2).[3] The Natufians who built it about 11,600 years ago were hunter-gatherers and seem to have remained such, as no clear evidence of farming has yet been found. They erected over 200 pillars of limestone in about twenty circles. In the middle of each circle were two taller pillars reaching up to 5.5 metres in height.

The limestone pillars were carved with abstract designs, as well as bas-relief images of a huge range of animals; the artists showed a particular penchant for vultures. Some of the pillars appear to be stylised human forms. Connected by low stone walls with benches, the resulting chambers have every sign of being designed for restricted performances. Smaller versions of the imagery on the pillars have been found in surrounding areas.

Archaeologists like to label buildings. Göbekli Tepe was tagged as a 'temple' early on in its rediscovered career. Evidence from oral cultures the world over shows that ceremonies, although they usually had a spiritual dimension, were not solely religious performances. The label 'temple' is misleading.

There have been some claims of astronomical alignments at Göbekli Tepe, but they have not been verified.[4] Most of the site has not yet been excavated. The resonance within the spaces are now well documented, however. Even gently striking the pillars with a hand will produce an infrasound resonance at 14 Hertz, a low sound that directly affects the body. Every chamber would have acted as a resonator for any music played there.[5]

From the same region of south-east Türkiye, and from about the same time as Göbekli Tepe, is a site known as Sayburç. In 2021, archaeologists excavated a building that they believe was used for communal gatherings. On a bench, carved into limestone, they found a wall relief, with human and animal figures. Eylem Özdoğan writes:

> It constitutes the earliest known depiction of a narrative 'scene', and reflects the complex relationship between humans, the natural world and the animal life that surrounded them during the transition to a sedentary lifestyle . . . This scene has the narrative integrity of both a theme and a story, in contrast to other contemporaneous images, and represents the most detailed depiction of a Neolithic 'story' found to date in the Near East, bringing us closer to the Neolithic people and their world.[6]

A few thousand years later, still in Türkiye, the communal nature of music and dancing was clearly represented on the walls of Çatalhöyük, one of the largest Neolithic settlements ever discovered. Up to 10,000 people were housed in single-room mudbrick homes, clustered together and entered through holes in the roof. One wall fresco shows a huge pig surrounded by many small human figures. There are men and women with drums, bowed stringed instruments, and what appear to be horns—the image depicts what has been described as an early orchestra.[7]

The amazing, almost unknown Stanton Drew

British Neolithic monuments are famous and spectacular. The mystery of how they were built has been understood by archaeologists for a long time. The mystery of *why* they were built is the one we can solve here.

What happens when you use a knowledge gene lens on an unfamiliar site? I offer you Stanton Drew as a test case. It is a massive Neolithic site that is pretty underwhelming to visit today, but it would have been spectacular in its time. In northern Somerset, about ten kilometres south of Bristol, Stanton Drew is mostly known through geophysical surveys (see Figures 9.1 and 9.2).[8]

When it was built, over 4500 years ago, Stanton Drew must have been an astonishing sight. It is a henge, a space surrounded by a ditch and bank. The enclosed circle was over 113 metres in diameter, ringed by a massive ditch, over six metres wide. Excavated large henge ditches, such as Stonehenge, Durrington Walls, Avebury and the Ring of Brodgar, all have flat bases, so we can assume that was the case at Stanton Drew as well. On the north-east side, a 50-metre entrance allowed people to cross the ditch. There is possibly another entrance on the other side of the henge.

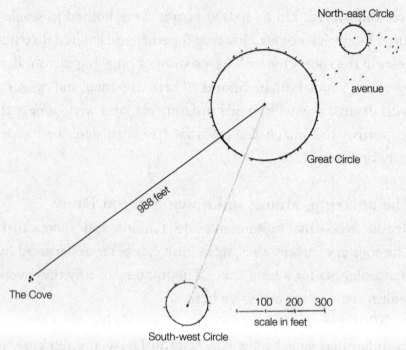

FIGURE 9.1 Plan view of Stanton Drew, redrawn from archaeological surveys.
IMAGE: LYNNE KELLY.

Just inside the ditch was a circle of standing stones just like those at stone circles all over Britain and western Europe. But the most spectacular part of Stanton Drew were the nine rings of posts. The huge tree trunks were set a metre apart, each post at least a metre across. These posts were carefully erected in orderly circles, beautifully spaced. That is no mean feat.

As well as this Great Circle, there were two smaller circles, another group of stones known as the Cove and two avenues. A number of possible astronomical alignments have been argued for the henge, the most strongly confirmed being with the solstices at midsummer sunrise and midwinter sunset, and the southern major and minor moonsets.

FIGURE 9.2 My interpretation of the archaeological reports of Stanton Drew, adapted from various sources.[9] IMAGE: LYNNE KELLY.

A distance from the rest of the monument, the South-west Circle is situated on top of a steep-sided hill; the centre 'is very flat and level, as if forming a deliberate platform', as archaeologist John Oswin describes it.[10] The archaeological report states:

[The] most striking thing . . . is how flat the site of the circle interior is, and it reinforces the opinion that the circle was positioned very deliberately to occupy the small plateau, with views northwards across the other circles towards the River Chew, westwards to the Cove, and across the valleys to the south and east.[11]

This smaller circle was perfectly designed for restricted small performances, while also making reference to the surrounding landscape.

In my reconstruction, I've turned the posts into decorated poles. Stanton Drew was a knowledge space. It satisfies all ten of the criteria discussed above.

The only alternative explanation I can find for the purpose of Stanton Drew is by the nature photographer, writer and presenter Rupert Soskin. In his film *Standing with Stones* (2020), Soskin suggests that the most likely explanation he can think of is that the site was designed for bloodsports, given the importance of hunting during the Neolithic.[12] The timbers represent a forest and the hunters slaying their prey could be viewed by the audience on the bank across the ditch. The ditch would stop the animals escaping. This explanation ignores any features other than the Great Circle, and fails to explain why the forest was so fanatically orderly. The huge posts would have to have been sliced at a height of a metre or so, he argues, to enable the audience to witness the gory spectacle.

I will stick to my knowledge space hypothesis, and hope I have convinced you. The time has come to apply the same logic to more famous monuments all over the world.

Going far north to Orkney

I have written about the Orkney monuments in more detail before, and regret that in a short section I can only offer you a glimpse of this extraordinary Neolithic landscape.[13] Off the north coast of Scotland, the Orkney archipelago consists of over 70 islands, which have supported human populations for over 6000 years.

Travelling to Orkney in 2010 was one of the most extraordinary experiences of my life. As the ferry approached the island, the tall red sandstone stack known as the Old Man of Hoy appeared to our right. I struggled to believe that I was finally there.

At the Standing Stones Hotel, a young lady with the most delightful accent booked us in. I could not contain my excitement. 'Where is the Ring of Brodgar stone circle? How do we get there?' She pointed out the window behind her. There, across the Loch of Stenness, was one of the most famous Neolithic stone circles in the world, and we could just walk to it. I calmed down, eventually.

One of the later Neolithic monuments on Orkney, the Ring of Brodgar (see Plate 9.3) was erected about 4500 years ago. The first stage of the monument was the ditch, just over 100 metres in diameter, about six metres wide and about three metres deep with a flat base. Even after nearly 5000 years of erosion, I found standing in the bottom of that ditch and staring up at the stones was an awe-inspiring experience.

The ditch is exactly the right height for performances that can be viewed from above, able to be well lit and out of the notorious Orkney wind. If necessary, it could be covered. As a performance space, it would have been superb.

Some leading archaeologists considered that the ditch acted as a moat, filled with water. Any decent hypothesis has to be able to be falsified, and this would really mess up mine. I was delighted when it was confirmed that the ditch would have been drained and dry almost all the time.[14]

A circle of about 60 stones was erected inside the ditch. Archaeologists estimate that it would have taken about 80,000 work hours to construct the ditch and stone circle. That is the equivalent of one person working 80,000 hours, 80 people working 1000 hours each or any other equivalent calculation you wish to apply. Any configuration represents an astonishing expenditure of energy.

The stones were perfectly spaced to be encountered singly. Archaeologists have reported that acoustic effects were produced within the circle by hand-clapping, vocalising and particularly by percussion.[15]

To travel from the hotel and the Ring of Brodgar, we needed to pass the smaller monument, the Stones of Stenness, and cross a narrow land bridge separating the Loch of Stenness from the Loch of Harray. It had been assumed that Orcadians had been walking over a natural hill between the monuments for around 5000 years. In 2002, that vision was upended when a geophysics survey found that there were Neolithic buildings beneath their feet. It is now referred to as the Ness of Brodgar, and the excavation of its buildings has been nothing less than spectacular.

The huge stones for the Stones of Stenness and Ring of Brodgar had been brought from distant locations, despite suitable stone being available much closer. These foreign stones would have acted as a direct reference to more distant landscape locations.

Rock art has been excavated at the Ness with abstract designs, as well as a carved stone ball.[16]

About 400 carved stone balls have been found at Scottish Neolithic sites, including Orkney.[17] Usually with a diameter of about seven centimetres, the stone spheres are the perfect size to be held in the palm of one hand and manipulated by the other as a memory device (see Figure 9.3).

Again, my hypothesis could be falsified. Some of the balls have well over a hundred protruding knobs on their surface. If these can't be read in a sequence, it really makes it hard to argue they act as a memory device. I held many such balls at the Hunterian collection at the University of Glasgow and at the National Museum of Scotland in Edinburgh. Every ball had an orientation and sequence to the knobs that was easy to identify.

Astronomical observations were possible at many Orkney sites, but most spectacularly in the famous chambered cairn of Maeshowe.[18] Within walking distance of the Stones of Stenness, the Ness of Brodgar and the Ring of Brodgar, Maeshowe is one of the largest passage cairns in Europe, Britain and Ireland. It is estimated it took over 100,000 work hours to build. The mound would have been over seven metres high,

FIGURE 9.3 A six-knobbed carved stone ball, photographed at the Hunterian, Glasgow (left). Photographed at the National Museum of Scotland, Edinburgh, were a many-knobbed stone ball (centre) and the elaborate Towie carved stone ball (right). PHOTOS: DAMIAN KELLY.

35 metres across and surrounded by a broad, flat space, which archaeologists consider was once a ceremonial site.[19] You can sense the contrast between the open public space as you queue outside to stoop and stumble through the narrow eleven-metre-long passage to the restricted space in the chamber. The passage was carefully aligned so the setting sun passes through the passage to strike the stones deep in the chamber for a few days each year around the winter solstice.

Although Maeshowe is often referred to as a tomb, the blocking stone for the entrance can only be closed from the inside—an unlikely act for the dead. In the two-metre-square chamber, a huge capstone, five metres above the floor, is held in place by spectacular curved drystone walls. There is no mortar, only brilliant stonework that has not let in a drop of rain in nearly 5000 years.

The Ness of Brodgar and its surrounding monuments were a central knowledge space serving the smaller knowledge centres dotted across the islands.

Over 80 smaller cairns across Orkney mimic the passage cairn design of Maeshowe, such as the Unstan passage cairn. Some contained human remains, but some also contained large numbers of animal bones. All would provide a perfect restricted performance space. The purpose of the cairns cannot be simply explained away as 'tombs', although honouring past Elders may well have been part of their purpose. Only a tiny proportion of the area's population was buried in the cairns.

Despite the impressive resonance of these sites, none of them could compare to the Dwarfie Stane (see Plate 9.4).

On the spectacular Orkney island of Hoy, the Dwarfie Stane is an 8.5-metre-long red sandstone block, hollowed out around 5000 years ago. Crawling inside, I could sit comfortably cross-legged. I chanted as low as I could to imitate a male Elder and the resonance was simply stunning. This would have been a sublime restricted location for a few people to sit and chant their knowledge.

Following the Neolithic came the Bronze Age. Individual burials in round barrows appeared across Orkney with grave goods, weapons and signs of individual wealth. As the knowledge elite lost their dominance, the resources available to them could no longer be justified. There were no new monuments. The stone circles and passage cairns were gradually abandoned as the transition to an agriculture-based economy was complete.

From Orkney to Ireland

Ireland offers a plethora of Neolithic monuments that display all the signs of ceremonial spaces providing knowledge centres.[20] Over 5000 years ago, the Irish started building passage cairns in the Boyne Valley of County Meath. They are almost universally referred to as tombs. Few bones were deposited over the millennium or so when the cairns were in use, so any burials represent only a tiny elite. Surrounding the restricted spaces within the cairns are stone and timber circles, henges and a proliferation of art and public performance spaces—exactly as you would expect if viewing the landscape in terms of knowledge systems.

Newgrange has a hideous white-quartz facade reflecting the architectural fads of the 1970s, probably one of the worst reconstruction makeovers known. Fortunately, the nearby earth-covered cairns of Knowth and Dowth offer a more realistic vision of what the landscape might have looked like during the Neolithic. There the quartzite stones have been laid out in front of the entrance, in the context in which they were found.

The mound at Newgrange is nearly 80 metres in diameter and would have originally been well over the current thirteen metres high. The deliberately flattened forecourt extended about 25 metres in front of the cairn. It would have served as a huge performance space, as would the large levelled area on top.[21]

Newgrange has a nineteen-metre passage, with artwork increasing as you approach the chamber. The largest capstone in the passage, only a few centimetres above your hunched body as you stumble in, is estimated to weigh 10,000 kilograms. The passage, and the chamber with its six-metre-high roof, are almost exactly as they were 5000 years ago. Remarkably like Orkney's Maeshowe, the sunlight on the winter solstice penetrates the narrow passage to reach right into the chamber. To complement the observations at Newgrange, the passage at Dowth is aligned on the midwinter sunset, hopefully ensuring the observers at least one clear view over the five or six days of solstice light beams, whatever the weather. Knowth isn't aligned, nor did it need to be.

Surrounding Newgrange are 97 carefully placed large kerbstones (see Plate 9.5), many heavily decorated, others of unusual shape. I could not imagine a better-designed set of mnemonic artworks. The stone and pit circles in the immediate vicinity would also have served as memory spaces. Similar stonework can be found at Knowth and Dowth.

There are over 600 decorated stones in the Boyne Valley, all incised with abstract art. Grooved ware, a decorated ceramic, has been found at Newgrange. It is also found at stone and timber circles all over Britain and Ireland, but is rarely found in settlements. Small, portable decorated stones have also been found among the rubble, many more probably lost in the reckless reconstruction of Newgrange.

The three large cairns are set on high points offering broad views of the surrounding landscape. The Neolithic builders also incorporated unworked stone brought from locations up to 80 kilometres away.

And just like on Orkney, smaller cairns can be found right across Ireland.

And so to England

Across England, there are many stone circles, timber circles and passage cairns just like those on Orkney. Interpretations seem to be highly skewed towards an obsession with death. The passage cairns are usually referred to as passage 'graves' or chambered 'tombs'. It seems fairly obvious that some of the knowledge elite would be buried there, or at least some of their bones, especially skulls, in homage to their revered knowledge. Many of the burials and cremations were in fact inserted sometime after the construction of the passage cairn, indicating that this was unlikely to be the initial purpose. Many of these monuments contained no human remains at all.[22]

People do a great deal more living than they do dying. I am not for a moment denying that death is a major consideration for all human societies, but food, shelter, family, social relationships, safety and (my personal bias) knowledge all outrank communal attention to dying in every contemporary culture I know.

Much as I would love to review every Neolithic site in England through the lens of the knowledge gene, that would soon become very repetitive: stones and timbers in sequential settings, decorated objects, astronomical alignments, public and restricted ceremonial spaces and artwork. So I will just briefly visit the main sites—Avebury and the broad Stonehenge landscape. I have written about these in depth previously, including a detailed academic treatment in *Knowledge and Power in Prehistoric Societies*.[23]

Over a thousand kilometres south of Orkney, Avebury is the largest stone circle in Britain. Around 5600 years ago, long before the main henge was constructed, the early Neolithic people spent an estimated 63,000 work hours enclosing over eight hectares within three concentric 'causewayed' ditches at Windmill Hill. The causeways crossed flat-bottomed ditches. Incised chalk balls have been found there. The later monuments include the passage cairn of West Kennet Long Barrow, which again offered public and restricted performance spaces. Archaeologists estimate that activities continued in West Kennet Long Barrow for a thousand years after the last burial.

The most stunning of the monuments in the Avebury landscape is the massive henge. At 420 metres across, it is about four times the diameter of Stonehenge. Building started around the same time as the first stone circle at Stonehenge was commenced, about 5000 years ago (see Figure 9.4). After five millennia of erosion, the maximum depth of the ditch is now far short of the original ten metres. Nevertheless, if you go down in the ditch and look up, the stones become dwarfed (see Figure 9.5). This thousand-metre-long ditch was cut into the chalk on an inconceivable scale.

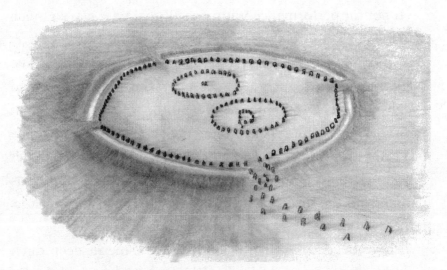

FIGURE 9.4 Avebury henge plan. IMAGE: LYNNE KELLY.

FIGURE 9.5 Illustration showing the scale of the Avebury ditch against the stones. IMAGE: LYNNE KELLY.

There must have been a crucial reason to justify the estimated 3,000,000 work hours to dig such a ditch with only deer antlers and stone tools. The excavating archaeologists commented on the significant effort the Neolithic users had made to keep the base totally flat.[24]

If not for performance within the ditch, why was so much care taken to ensure a flat base? By comparison, the later Iron Age ditches were defensive, with a V-shaped profile. The Neolithic avenues, such as at Avebury and Stonehenge, had rough-cut ditches with no specific profile. Flat bases were cut in the dazzling white chalk, where the performances would be protected from the weather. With the light from the torches reflecting from the towering white sides, and the sound resonating from the stone walls, this must have been theatre at its most awesome.

I have begged Damian to dig me a ten-metre-deep ditch into the stone in our backyard. Just a segment would do. So far he has refused.

The stone circle at Avebury consisted of up to 100 huge stones erected just inside the ditch, all retaining their natural form. There are two small inner circles, providing restricted spaces, including a stone arrangement known as The Cove, which is aligned on the solstices.

On nearby Overton Hill, the Neolithic Britons erected six concentric timber circles and two of stone, which is now known as The Sanctuary. They made their own hill, Silbury Hill, a completely human-made mound with a diameter of 165 metres. On top, 40 metres above the ground, is a flat top that stretches 30 metres across. At that height, it is a restricted performance space. There are no burials or original tunnels. Silbury Hill is exactly what you see: a mound of chalk surrounded by a wide public space enclosed by a six-metre-deep ditch.

The whole Avebury landscape area was a massive, dynamic gathering place, constantly changing over the millennia.

Every stage reflected exactly what the knowledge elite required to serve the memory needs for all the communities whose members ventured there.

And now to Stonehenge

Tlingit Elder and storyteller David Kanosh visited Stonehenge in 1982. He described the experience:

> In 1982, we went to the UK. We went to Stonehenge and I loved it.
>
> Our tour guide unsuccessfully tried to get me to calm down and stay with the group. I was running from stone to stone. I touched each stone. I hugged each stone. I wondered what part of what stories each stone helped the storyteller remember.
>
> My grandfather told me if these stones were used for storytelling, the storytellers who used them were long gone.
>
> I walked around the outside of Stonehenge. I turned to look outwards in every direction as I went. The tour guide asked me what I was doing.
>
> 'There's gotta be more to this than what we see. There's some things out there that I think help tell a story. There has to be something out there,' I replied. The tour guide shook his head. He herded us back to our transport and we left.
>
> Scotland, Ireland and over to Brittany in France, each had rock formations which had my family in awe. I quietly wondered what stories these rock formations helped a story-teller recount. They certainly dwarfed any rock formation we had in south-east Alaska.

Although I'll only give a brief overview of Stonehenge here, it should be sufficient to show that the monument was designed, built and constantly adapted to suit the knowledge needs of Neolithic Britons.

I'm sure it is the final structure that is conjured up in your mind when you read the word *Stonehenge*. But the initial monument looked nothing like that. It is estimated that about 30,000,000 work hours were spent on the monument over its many phases (see Figure 9.6).

The earliest known monument in the Stonehenge land-scape was a line of massive posts, dated to 4000 years before the first stone circle. Then there were long enclosed stretches of land, such as the three-kilometre-long Stonehenge Cursus. To the people of ancient England, this had long been a sacred site when they started building henges. Stonehenge was one of the first to be built, and one of the last to be abandoned.

Around 5000 years ago, the first circle at Stonehenge was a simple one about 100 metres across. Posts or stones were erected in 56 pits with about six metres between each. It is generally believed that these holes held bluestones brought from Wales. It must have looked pretty much like the Ring of Brodgar.

The ditch was cut into about 60 segments of varying depths and lengths, using only deer antlers and stone tools as digging implements. Each segment had a flat base. The size of each segment would have been determined by the needs of the particular ceremony being performed in the annual cycle. The depths of the segments averaged about two metres. This would allow a temporary cover of skins to be added to protect performers from the notorious British weather. Over

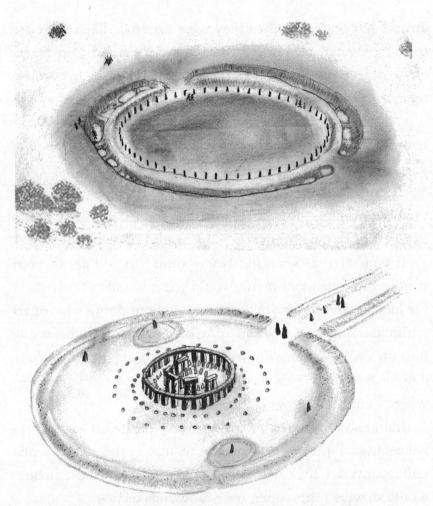

FIGURE 9.6 The initial stage of Stonehenge dates to about 5000 years ago. The sarsens were erected at Stonehenge about 4500 years ago. IMAGE: LYNNE KELLY.

the length of the use of the henge, some of Stonehenge's ditch segments were left to erode while others were maintained, as would be expected as the ceremonial needs of the population changed during the Neolithic transition.

The builders had deliberately placed objects in the Stonehenge ditch, including animal bones that were already between

70 and 420 years old when they were inserted. Then there are the Stonehenge plaques, abstract designs carved into heavy chalk objects, a perfect size to be held and used as portable memory prompts. Archaeologists have suggested that symbols were probably carved into the ditch walls, or that the walls had been painted, but the art has been lost to thousands of years of erosion.

There is very little indication of art at Stonehenge other than the plaques—something that has concerned me; I would expect a great deal more art. But I think I have an answer.

If you were a Neolithic Brit wanting to add art to your wonderful stone monument, would you head off to find ochre? Or maybe, given that you're thinking about doing a bit of art while standing on a chalk plain, where ditches have been dug into the chalk, leaving numerous chunks lying around, where designs have been carved on lumps of chalk . . . can you see where I am heading with this?

Indigenous cultures don't necessarily create art for permanence. In fact, they tend to constantly update the images when telling stories. Chalk would be perfect. None of those images would survive in the open over thousands of years.

Around 4500 years ago, 500 years into the life of Stonehenge, the Neolithic Britons dragged the huge stones, known as sarsens, to the site from 25 kilometres away, near Marlborough.[25] They erected the five towering trilithons at the centre and surrounded them with 30 upright sarsens in a tight circle. These were topped with a carefully worked interlocking ring of lintels—an incredible achievement (see Figure 9.7).

The precious bluestones were moved from their wide circle

FIGURE 9.7 The sarsens and bluestones in the centre of the Stonehenge monument as it appeared in the final stage of use about 4500 years ago. IMAGE: LYNNE KELLY.

into the centre, arranged in the order they had been before and protected from public gaze by the massive sarsens. More bluestones were added to the central arrangements. Stonehenge had been adapted to serve as a much more restricted site.

Various researchers have explored the acoustic enhancement at Stonehenge in its final stage. The resonance and amplification of drums and other percussion instruments would have been sensational. Any performance would have produced substantial sound effects within the space enclosed by the sarsens, yet very little of the sound would have escaped the restricted theatre.

The second delivery of bluestones, which were then shaped, possibly acted as a set of sounding stones when they were struck.[26] Wouldn't that smoothing also make them better blackboards?

That Stonehenge offers solstice alignments is well known. The builders even shifted Stonehenge's axis by a few degrees when they erected the sarsens to improve the accuracy in aligning with the midsummer sunrise and midwinter sunset. But archaeologists now believe that Stonehenge was a far more complex calendar than just aligning with the solstices.[27]

By adding the sarsens, Stonehenge had become a restricted space. My hypothesis requires both public and restricted performance spaces. But at the same time Stonehenge was being upgraded, a huge ceremonial complex was being constructed within walking distance at Durrington Walls.

There are a lot of different ceremonial spaces at Durrington Walls. The size of the henge is overwhelming. The ditch is over a kilometre in length, enclosing a circular area 400 metres in diameter. It is divided into approximately 22 segments, averaging 40 metres in length and up to ten metres in width—perfect for performances. Each flat-floored segment was the required size, larger or smaller depending on the size of the group entrusted with a particular ceremony. Even more impressive, some segments are over five metres deep. Every basketful of chalk had to be chipped away using only the antler of a deer and then hauled by hand to the bank.

In the ditch bank, archaeologists have found over 50 mega-liths in a 300-metre row. Within the centre area, there are two huge timber circles, one with six concentric circles of posts, which would have made these timber circles fairly restricted but also an amazing site if each of those poles were carved and decorated, as logic tells us they would have been.

A short 70 metres from the henge is an even more restricted site known as Woodhenge, named for the 168 posts estimated to have been over seven metres high. Six concentric circles, all within the hundred-metre diameter, were contained by a flat-based ditch and bank.

Stonehenge and Woodhenge aligned with the midwinter sunset and midsummer sunrise. Durrington Walls was also aligned on the midsummer sunrise, while the Durrington South Circle was aligned to the midwinter sunrise. And grooved ware has been found, the decorated pottery so closely associated with the henges, and so perfect as a memory device.[28]

If the purpose of Durrington Walls was not to be a knowledge space, it is very hard to understand why so much energy was expended on its construction. There is no evidence of anything being grown or produced that might have been traded. What was being traded was knowledge.[29]

There are many more features in the Stonehenge/Durrington Walls landscape—too many to mention here, but the pattern is clear. There are large central knowledge spaces with both public and restricted performance spaces, aligned on the solstices. There are stones, each varying, and tall timbers, each able to be carved and painted. There are replicas of these massive memory spaces dotted around the entire British and Irish landscapes in the thousand or so smaller circles.[30] There is art. There is resonance for music and chanting, and ways to ensure performances could proceed whatever the weather. Memory spaces were being expanded everywhere, exactly as would be needed as the population grew and diversified.

Was knowledge their only purpose? I very much doubt it. There are no ethnographic examples of gathering places that exist purely for knowledge. Gatherings also served as social events to meet prospective partners, for initiation and other ceremonies, to trade material goods and for having parties.

As power moved from a knowledge elite to those who exploited their warriors and wealth, burial mounds appeared in the archaeological record, rich with grave goods. Eventually, the monuments in the Stonehenge landscape were abandoned. Magnificent as they were, they could no longer demand the expenditure of energy necessary to serve their memory masters.

That did not mean that the knowledge experts disappeared from the populations. They just moved down the power chain a little bit. Around 2000 years ago, Julius Caesar and other classical writers wrote that oral specialists in Britain memorised vast numbers of verses during their twenty years of training, many of which dealt with celestial phenomena and calendars.[31] Education specialists—referred to as druids and priests—were essential and powerful, but no longer at the very top of what was now a clearly hierarchical society.

Stonehenges, Stonehenges everywhere

All over the world, spaces are tagged as somewhere's 'Stonehenge'. This might seem to be just for the media, just clickbait. But there is a revealing pattern when you compare these places. Across vast distances and thousands of years in time, there can have been no contact between the builders. The 'Stonehenges' are all open, non-domestic spaces with locations marked by stones. All are aligned on some astronomical event. All would have been perfect to serve as ceremonial knowledge spaces.

Dating from 7000 years ago is the 'Egyptian Stonehenge'. The megalithic stone circles at Nabta Playa align with the summer solstice. At around the same time, Rujm el-Hiri was built (see Plate 9.6). Tagged the 'Stonehenge of the Levant', it is a megalithic monument with concentric circles built up of stone. It is about sixteen kilometres east of the Sea of Galilee, and aligns on the summer solstice and the equinoxes.

There are a few claims to the title in Germany. The Goseck Circle was built about 7000 years ago. The circle was probably two concentric wooden walls surrounding a ceremonial space and ditch. The entrances align on the winter and summer solstices. The tag 'Stonehenge of Germany' is also given to Pömmelte ring sanctuary, which 4300 years ago would have been a series of timber rings. Some of the wood has signs of carved symbols.

I have visited 'Australia's Stonehenge' at Wurdi Youang, Victoria, with the Traditional Owners. Consisting of more than 100 basalt boulders, there are alignments with the solstices and equinoxes. It is not yet dated, but could be much older than its namesake.

I could continue listing sites that have been tagged as a 'Stonehenge' and give you a string of repetitive descriptions. I'll just name a few more so you can look them up and assess the patterns for yourself:

- The Blaauboschkraal stone ruins in Mpumalanga, South Africa.
- The megalithic site at the village of Byse, in Karnataka, India.
- Stone circles located in the Flaming Mountains in Turpan, in the Gobi Desert, China.
- Stone circles from Ōyu and Isedotai, in northern Japan.

- The megalithic *taulas* of Menorca, in the Spanish Balearic islands.
- Megalithic circles at the Dolmen of Guadalperal, Spain.
- The Blomsholm stone ships and Ale's Stones, among many 'stone ships' in Sweden.
- The Giant's Churches of Kastelli and Kettukangas, among others in Finland.
- *Mustatils* in Saudi Arabia, of which there are over 1000.
- *Saywas*, in the Atacama Desert, Chile, bordering the Inca Trail.
- *Ha'amonga 'a Maui*, in Tonga, known as the 'Stonehenge of the Pacific'.
- The numerous Senegambian stone circles found in The Gambia and in central Senegal.
- The megalithic complex known as the Cromlech of the Almendres in Portugal.

Even more Neolithic monuments fit the pattern

The Neolithic monuments of Malta have public and restricted performance spaces and some of the most extraordinary acoustic enhancements known anywhere in the ancient or modern world. There are the two Ġgantija temples on the Maltese island of Gozo, the Mnajdra megalithic complex and the Hypogeum of Ħal-Saflieni. I would so love to visit them.

The most extraordinary megalithic site that I have visited was in France.[32] The French region of Brittany offers monumental archaeology dating back to over 6000 years ago. There are around 250 decorated stones across 75 sites, many very like the passage cairns of Ireland. There are over 3000 standing stones, known by the French term *menhirs*. Some stand alone, some are in circles, but most are in the extraordinarily long sets of parallel rows at Carnac (see Figure 9.8).

Walking the three main groups of rows at Carnac—Le Ménec, Kermario and Kerlescan—was an extraordinary experience. They just go on and on and on, yet each stone stands alone, perfectly distanced from its neighbour.

Nearby, on the shore of the Gulf of Morbihan, is the most highly decorated passage cairn of all, Gavrinis, with its own two stone circles, the stones of Er Lannic. Not far away is what was the largest standing stone in Europe, the Grand Menhir Brisé at Locmariaquer. Once a single standing stone,

Stone Avenues Carnac, Brittany.

Fig. 3. Plan of portion of the Avenue at Carnac, Brittany.

FIGURE 9.8 The rows of standing stones at Carnac, France, in 1870. From J.B. Waring, *Stone Monuments, Tumuli and Ornament of Remote Ages*, John B. Day, London, 1870, p. 43.

the Grand Menhir measured 21 metres long, rising almost 20 metres above ground. It would tower over even the tallest of the sarsens of Stonehenge, which stand a mere 6.7 metres above ground. There must have been an incredibly important reason to raise this 300-tonne stone, because it would have required hundreds, if not thousands, of people to cut it from its location in the bedrock, drag it several kilometres to Locmariaquer and then stand it upright.[33]

Along with the standing stones, the Neolithic landscape of Brittany was enriched with mounds, referred to as tumulus monuments. There were public and restricted spaces in abundance. Art adorns the walls of the cairns, which have superb acoustic properties. Music, art and connection to place in non-domestic ceremonial spaces—the pattern continues.

Although the stones are in their natural state now, recent research indicates that they were very likely painted. Archaeologists led by Primitiva Bueno Ramírez, from the University of Alcalá in Spain, researched traces of pigments on megaliths in southern Europe dating from 8000 to 3000 years ago. They found that many megaliths were coloured. Pigments were used to create geometric and representational motifs.[34] What we see now are silent, bland versions of the monuments, so sterile when compared with what they must have been when their communities were bringing them to life with colour and theatre.

At Nazca, they didn't have any megalithic stones

If you want standing stones, then you need boulders. But standing stones aren't the only way to create astounding memory spaces of monumental dimensions. In the foothills of the Andes in Peru is a plain known as the Pampa de San José. There, the Nazca people started moving millions of small, dark stones. Over 2000 years ago, they created pale paths that mapped out huge animal glyphs, geometric shapes and very long lines.[35]

There's a bird with a neck that zigzags for half its 300-metre length, and a monkey with a coiled tail 300 metres in diameter. There are human figures and animals that are part plant. There are hummingbirds, pelicans, foxes and condors, an insect and a glorious spider. A 285-metre-long pelican dominates the animals, but the creatures are outnumbered by the spirals, zigzags, triangles, trapezoids and long, straight lines totalling well over a thousand kilometres of paths (see Figure 9.9).

The Nazca Lines contain over a thousand figures, many with astronomical alignments.[36] The lines were also about water, critical in the desert, pointing to the underground resources and rivers in an environment with no rain and dependent on flow from the Andes.[37]

The aspect that convinces me these are memory spaces is that no matter how complex the figure, no line crosses another. All figures can be traced using a single continuous line. These images were created to be walked. The archaeology indicates that several of the shapes on the 200-square-kilometre desert plateau were enclosed performance spaces. Others were glyphs intended to be walked.

As the Nazca stepped onto the glyph, chanting as they took the familiar path, they could do so knowing they would never cross the path of another walker. The Nazca had exploited their innate knowledge gene skills to make art, while also

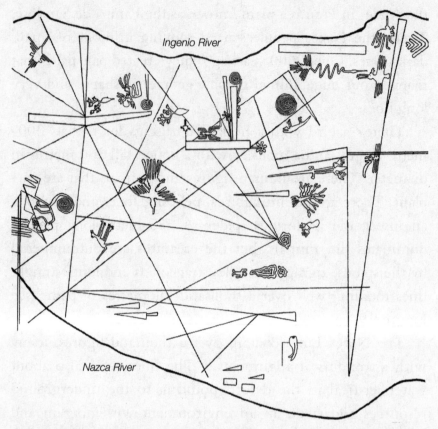

FIGURE 9.9 Part of the Nazca Lines, including the animal glyphs, lines and trapezoids. The monkey is on the left. IMAGE: LYNNE KELLY.

displaying extraordinary spatial skills; when walked with a paced rhythm, this created a memory space that would have been phenomenally effective. It must have been a profound experience to walk those glyphs and lines or participate in performances within the enclosed trapezoids.

There is a misconception that the shapes needed to be viewed from the air. That's because that is the way the lines are presented to us. The Nazca could see the animals from nearby hills; the lines needed no such view. Although from the air the designs appear to overlap, on the ground, the Nazca would walk the design in use at the time. The Pampa de San José was an ever-changing canvas, with new designs overwriting those that had lost relevance.

As would be expected with animal forms representing oral tradition, the same characters are represented in art on other media. In the Brooklyn Museum, New York, I was delighted to see the monkey on a Nazca jar (see Plate 9.7).

While rushing through the museum to meet two of the curators for lunch, I glimpsed another fascinating object on display in a small gallery. It was a tapestry I had read about but never dared hope that I would see. Known as 'the Paracas Textile', it is an extraordinarily complex mantle or cloak, measuring 170.8 centimetres by 84.5 centimetres, with 90 needle-knitted figures around the perimeter (see Plate 9.8).[38] Considered a ceremonial mantle, it contains many images of the plants and animals important to the Nazca, to whom it is now attributed.

The Paracas Textile is breathtaking. I cannot imagine a more perfectly designed or more beautiful portable memory device. I was late to lunch.

Around 1900 years ago, the Nazca built a massive ceremonial centre on the bank of the Nazca River at Cahuachi. At the focal point of many of the straight lines, Cahuachi had mounds, step pyramids, plazas and a small room with twelve freestanding posts in neat rows. This was not a residential city, yet a huge amount of highly decorated pottery has been found there.

The Nazca had public and restricted memory spaces, music and art in locations maintained away from residences: the monumental pattern continues.

Monuments across the vast Pacific Ocean

I have been focusing on cultures that structure their knowledge systems on landscape locations. Polynesian cultures structure their knowledge systems on their ancestors, biological or mythological, while still taking advantage of monumental spaces.

What we refer to as 'Polynesia' is a huge area of the Pacific Ocean, best imagined as everything within a triangle with Hawaii at the top, Aotearoa (New Zealand) at the bottom and Rapa Nui (Easter Island) way out to the east. In the middle of Polynesia are the Cook Islands, the largest being Rarotonga. A specialist on Rarotongan oral tradition, Matthew Campbell, wrote:

> [T]he building of monuments is a near universal activity, and their purpose is also universally bound to memory. In most preliterate societies memory is preserved in oral tradition and genealogy, and such societies will build monuments with genealogical memorial practices in mind, as the Raratongans did with the Ara Metua and the marae.[39]

The Ara Metua is a ceremonial road, while the *marae* are communal sacred buildings. Carved figures act as memory cues for complex genealogies.

Campbell describes the Polynesian *marae* as serving a very similar purpose to the Hawaiian sacred places known as *heiau*, and the *ahu* platforms with their *moai* statues on Rapa Nui. The Aotearoan Māori also build *marae*. As we'll see, the knowledge practices of these cultures are very similar, despite the physical distance between them. Given the legendary navigation skills of Pacific islanders, they were not as isolated as a map of the enormous Pacific Ocean might suggest.

Let's start in Hawaii, on the small ceremonial island of Mokumanamana (Necker Island). At the Papahānaumokuākea Marine National Monument, there are 33 ceremonial *heiau* with rectangular platforms, forecourts and standing stones.[40] Portable stone images with human form, known as *ki'i pōhaku*, were found in the *heiau*.[41] Elsewhere in Hawaii, large wooden *ki'i* statues stood on platforms.[42]

Polynesians navigated their way to a new home on Aotearoa around 800 years ago. The knowledge system maintained by the Māori is structured by genealogies in much more complex formats than simple family trees. The Māori system of genealogy, *whakapapa*, relates everyone to their foundation ancestors and interweaves all living things and all landscape features.

Māori writer Bradford Haami describes *whakapapa* as an essential framework for memorising a chronological sequence of events, with the ancestors being the source of knowledge.[43] Typically, Polynesian cultures recall an amazing 25 generations back in time. Beyond that, biological and mythological ancestors become interwoven in the knowledge system. Haami explains that a large number of ancestors are required for the knowledge keepers to have sufficient stories to store enough information. But too many ancestors becomes unwieldy.

Traditional knowledge, *Mātauranga Māori*, is maintained through both public and restricted performances at the *marae*. Haami talks about Māori ancestral houses as being like learning centres or libraries, with the carvings being the equivalent of books, encoding the knowledge through memorialising the ancestors. Each ancestor has an associated oral tradition.

On Aotearoa, they carved their representations of the ancestors in wood. On Rapa Nui, they used stone. Rapa Nui is a monumental island.[44]

The new arrivals on Rapa Nui took advantage of the readily available soft volcanic stone to carve over a thousand *moai*, each one different from all the others. The majority of statues were quarried from the volcanic crater at Rano Raraku and transported around the island. They were never placed in domestic situations. They were for ceremony.

More than 200 *moai* stood on the platforms known as *ahu*, facing the volcano crater with their backs to the sea. Some of the inland platforms seem to be aligned with the winter solstice sunrise. Fifteen *moai* once stood on the great Ahu Tongariki (see Plate 9.9). In front was up to 75 metres of smoothed stonework, and a further 50 metres or so of levelled ground. This was a superb public performance space.

British archaeologist and anthropologist Katherine Routledge stayed on Rapa Nui for seventeen months from 1914. She wrote:

> Looked at from the landward side, we may, therefore, conceive an ahu as a vast theatre stage, of which the floor runs gradually upwards from the footlights. The back of the stage, which is thus the highest part, is occupied by a great terrace, on which are set up in line the giant images, each one well separated from his neighbour, and all facing the spectator.[45]

Many more *moai* are along the roadways or on the side of the quarry mountain, in what Routledge described as 'magnificent avenues'.[46] Some are still within the crater, even today considered a restricted place by the Indigenous population.

The captains of the canoes landing on Rapa Nui would have been highly trained Pacific navigators relying on their famed memory systems. The newly arrived Polynesians began carving statues that represented the ancestors who populated their memorised stories.

In 1862, slave traders raided the island, annihilating the adult population. Only fifteen returned to the island, bringing smallpox with them. By 1877 there were only 111 people left on Rapa Nui. With the senior men and women gone and the remaining population decimated by disease, invading farmers were able to commandeer the traditional lands.

Missionaries converted the survivors to Christianity. The oral tradition could not possibly survive. The *moai* had been toppled from their platforms as their power had clearly failed.

The Indigenous Rapa Nui responded with a new cult of the Birdman.

Many art forms, including figurines, are indigenous to Rapa Nui. Most intriguing are the wooden tablets, of which a few dozen survived the orders of the missionaries for their destruction. Inscribed on the tablets is *rongorongo*, long sequences of abstract glyphs, along with images of people, birds, marine animals and plants (see Figure 9.10). There are astronomical bodies, boats and ships, tools, weapons and ornaments. Translation has proved challenging, even though the tablets are only about 100 years old. Is *rongorongo* a written script yet to be deciphered? Katherine Routledge believed that it was proto-writing that acted as a memory aid for the island's history and mythology.

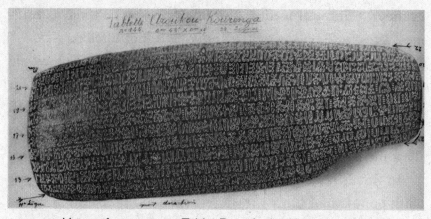

FIGURE 9.10 Verso of *rongorongo* Tablet B, or Aruku-Kurenga; dating from before 1860, it is 43 centimetres long and 16 centimetres wide, with ten lines on the front and twelve on the back—a total of 1135 signs.
PHOTO: STÉPHEN-CHARLES CHAUVET (CREATIVE COMMONS LICENCE CC BY-SA 3.0).

North America abounds with mnemonic mounds

All over the Americas are monuments designed as knowledge theatres. I can only mention a few but the same pattern can be seen. Indigenous Americans took superb advantage of their knowledge gene skills.

There is not much to see on the surface at the mound-building site of Poverty Point, in Louisiana, yet the site is an extraordinary example of the fact that hunter-gatherer-fisher cultures are perfectly capable of building sensational monuments.

I have previously written in much more detail about Poverty Point, a site I visited in 2009, which was a truly fascinating experience.[47] There is, famously, no stone bigger than a grain of sand in the region of the extensive Poverty Point culture, so their performance spaces were bounded and raised with mounds of earth. Some mounds were used for burials

and residences, but many were primarily non-domestic ceremonial sites.

When I visited Poverty Point, archaeologist Dr Diana
Greenlee introduced me to the elder statesman for the archaeological community, Dr Joe Saunders, who offered to take
me to the oldest of the mound-building sites. Watson Brake
was built about 5500 years ago on the Ouachita River. Eleven
mounds surround the central public plaza. The top of the
largest mound, even after millennia of erosion, would have
been a perfect restricted performance space. Traipsing such
ancient mounds in the rain was astounding.

The better-known site of Poverty Point was built nearly
2000 years later, starting around 3700 years ago.[48] Six semi-
circular earthen rings surround the plaza, with some spaces,
such as Mound A, situated high above the public arena. None
of the mounds contains burials. Hundreds of posts had been
erected on the ridges.

Over time, 30 post circles were built and replaced in the
plaza, with similar proportions to those in the British Neolithic
and those we'll see at the later Indigenous American site of
Cahokia. There are numerous decorated portable objects at
Poverty Point—some are human and animal figurines, but
most are abstract designs. There are many smaller sites along
the Mississippi, replicating the Poverty Point cultural artefacts
in a widespread culture.

A huge amount of stone was brought to Poverty Point
from as far as 2400 kilometres away. As always, if there is trade
coming in and nothing material going out, then they must
have been trading knowledge. I couldn't help imagining the
knowledge performers contending, as I did, with the dense

swarms of oversized mosquitoes. I do hope their knowledge system included repellent.

Indigenous Americans were superb mound-builders. There are the flat-topped pyramid mounds of the Coles Creek culture in the Lower Mississippi states of Arkansas, Louisiana and Mississippi. Effigy mounds take the shape of animals, such as the Great Serpent Mound and the Alligator Effigy Mound in Ohio.

The Woodlands cultures—including the Adena, Hopewell and Fort Ancient peoples—lived on the Great Plains of North America. Among the Adena's many achievements is the Mount Horeb Earthwork in Kentucky, dating back about 2500 years. The earthwork is described as a henge, the interior being bounded by a ditch and a bank. Pre-dating the entire structure, which is nearly 100 metres across, was a post circle. Researchers consider this to have been a ceremonial space.[49]

Like their predecessors on the Great Plains, the Hopewell built platform mounds, ceremonial enclosures and post circles.[50] At the Stubbs Earthworks in Ohio, the Great Post Circle held 151 posts in a circle that was 75 metres in diameter. Nearby, the Moorehead Circle boasted three concentric rings of posts, the outer ring being about 60 metres in diameter. Like so many post circles in North America, the Moorehead Circle has been nicknamed 'Woodhenge', making reference to the British Neolithic site at Durrington Walls.

And then there's the most complex of all: the Mississippian culture cities at Moundville in Alabama and at Cahokia in Illinois.[51]

At the largest of these cities was yet another 'Woodhenge'. Occupied from around 1000 years ago, Cahokia is thought to have supported a population of up to 20,000 people. At least 100 mounds surrounded the massive Grand Plaza. The largest, Monks Mound, was a flat-topped pyramid that grew in stages to reach 30 metres in height. Along with the ceremonial mounds, other mounds appear to have been for burials while yet another was a copper workshop.[52] Cities of this size are much more than just memory spaces.

Cahokia's Woodhenge (see Plate 9.10) varied over its life, being constantly rebuilt. It was usually around 120 metres in diameter with up to 60 posts, aligned on the solstices and equinoxes. Red ochre in the pits indicates that the posts were probably painted. I suspect they were very like the Indigenous American totem poles found in the north-west of the continent. There are many superb examples of artworks as well, both abstract and representational.

Dr Julie Zimmermann Holt, Professor of Anthropology at Southern Illinois University, argues that Cahokia could be considered the centre of a theatre state.[53] She sees Cahokia, as do most researchers, as an economic and political centre, but also emphasises its role as a centre of ritual. She proposes that Cahokia's power lay more in its ceremonies than in its armies. People came because they wanted to be part of the drama.

The extraordinary Chaco Canyon

Much as I would love to explore the hundreds of other sites I have viewed through the knowledge gene lens, it is beyond the scope of this chapter. But I cannot possibly leave out the site that had more impact on me than any other. I fell in love with Chaco Canyon, in New Mexico, and spent much time there just staring at the colours, appreciating the silence and being dazzled by the awesome archaeological remains. I have written in detail about the Pueblo, Ancestral Pueblo and the astounding Chaco Canyon before; over a decade later, my feelings of awe have not abated.[54]

Around the same time as Cahokia flourished, about a thousand years ago, the Ancestral Pueblo had turned Chaco Canyon into a monumental space the like of which exists nowhere else. In the northern New Mexican desert stand the remains of more than a dozen 'Great Houses'. Hundreds of rooms in buildings up to four storeys high have been wonderfully preserved by the dry atmosphere of the desert. What I found most intriguing is that there is little archae-ological evidence of anyone living in them, although there is evidence of a small permanent population, a powerful elite who relied on neither coercion nor wealth to bring in the labour.

Within each Great House were highly restricted circular ceremonial rooms, known through contemporary Pueblo cultures as *kivas*. The largest of the Great Houses, Pueblo Bonito (see Plate 9.11), once stood at least three storeys high and enclosed 47 *kivas* among over 650 rooms surrounding a central plaza. All the Great Houses in Chaco Canyon followed a similar pattern, with over 200 smaller versions, known as Chacoan Outliers, across a region of 100,000 square kilometres.

One of the contemporary Pueblo language groups, the Hopi, refer to Chaco Canyon as *Yupköyvi*, which means 'a place where knowledge was to be shared'.[55]

The Great Houses were usually located in exposed promi-nent locations, which enabled a direct view of significant landforms marked with shrines and rock art. Within the

Great Houses, a vast array of art objects were found, many in restricted distributions.

The Ancestral Puebloans cleared over 300 kilometres of 'roads', averaging about nine metres wide, leading in and out of the canyon. There were no wheeled vehicles and no beasts of burden. The roads took longer, steeper routes than was necessary, and included both staircases and long gaps. The only logical conclusion is that these roads were memory trails connecting the most dramatic landforms, which were often marked by small stone cairns. This resonates with the Zuni Pueblo pilgrimage trail, which consists of hundreds of kilometres connecting the Zuni Pueblo to the Bandelier National Monument (described in Chapter 4). The contemporary Pueblo recite the names of the shrines in chants that still guide their pilgrimages.

Isolated *kivas*, such as the huge Casa Rinconada, were built as Chaco Canyon reached its zenith. Theatre was paramount. A screened area was entered via a subterranean passageway, which then led to the performance space. Regular niches around the walls would have served as memory points. There were large pits, probably used for foot drums. In the reconstructed *kiva* at one of the Chacoan Outliers, mistakenly named Aztec Ruins, I was able to experience extraordinary acoustic effects. The Ancestral Pueblo were using their musical and visuospatial skills with flair.

In a few rooms at Pueblo Bonito there were cylindrical black-on-white ceramic vessels referred to as 'cylinder jars'. Only 210 have ever been found. These jars were constantly redecorated, each with a unique pattern. The only place in Chaco Canyon archaeologists have found traces of chocolate, which must have come from thousands of kilometres to the south, is in these jars. Intriguingly, the same patterns adorn the decorated wooden flutes found in the Great Houses.

Contemporary Puebloan sun-watchers are renowned for their complex astronomical knowledge: they maintain accurate records of the sun, stars and planets. The Ancestral Pueblo built astronomical alignments into their buildings, too, even tracking the 18.6-year lunar standstill cycle. They engraved complex artworks high in the mesas to track the solstices.

A great volume of valuable material goods was brought from great distances to Chaco, including vast amounts of turquoise. Also entering the site was wood, food and labour. Nothing seems to have been manufactured there. The only way to pay for these goods and services would be with something ephemeral: knowledge.

I must leave it to the reader to test the insights that the knowledge gene lens offers for the many other sites across North America. I highly recommend that you take the time to consider whether these cultures created their ceremonial sites primarily to retain their precious knowledge systems.

What do these ancient monuments have to tell us about today?

People in the past didn't outsource their knowledge to dusty books on library shelves, or to some corner of the internet, never to be visited again. Much as it pains me as an author to say it, neat words on paper or a screen are not enough.

As I immerse myself in my memory palaces, I am glimpsing what it means to feel embedded in your knowledge system, grasping it with all your senses. It feels like my knowledge is alive and is being kept alive. It reminds me of the way Margo Neale talks about singing up Country. It just feels so natural, and now I know why. I am tapping into the potential the knowledge gene offers that I have never used properly before. And I feel this is just the start.

There is a global pattern: stone, timber and earthen circles; public and restricted spaces; decorated walls, rocks and enigmatic objects; astronomical alignments; enhanced acoustics; and music. These all worked in combination to enable extraordinary knowledge skills. We know stories have survived from over 10,000 years ago from south-eastern Alaska and from Australia, pre-dating these monuments by thousands of years.

As we have seen, like their contemporaries the world over, the Indigenous Australians and Alaskans built monuments as knowledge spaces to aid the performance and maintenance of their precious stories.

Despite the diversity of human cultures across the Americas, Africa, Asia, Europe and Oceania, the similarities in the knowledge technologies are striking. There is a clear reason for this: the human NF1 allele is driving knowledge systems in all of us.

CHAPTER 10

The impact of literacy on our knowledge gene skills

For the vast majority of human existence, there was no writing. Our ancestors used a combination of three communication systems, superbly optimised to work together: language, music and art. The previous chapters have shown that all of these skills were ancient, innate, universal, uniquely human and so valuable that evolution ensured that the genetic mutations which enhanced them spread through us all, despite any vulnerability they might cause in a small proportion of people.

And then came writing.

What happened to our NF1-granted skills when literacy arrived? Everywhere, writing sidelined music, art and our association of knowledge with spaces. As literacy took hold, humans continued to build monuments. They just aren't as mysterious anymore, because we have written records to tell us what they are. So we started calling them temples, palaces and cathedrals. Writing and scribes gradually took over the role of storing knowledge.

What is writing?

There are many definitions of writing, but I am going to take a very narrow and pragmatic one here. I define writing as the creation of signs that can be read by an independent person who will speak, word for word, exactly what the writer intended. Every utterance can be represented—and that requires symbols for sounds.

Inventing writing is not a simple thing to do. It's an incredible cognitive leap.

Writing appears to have started quite independently in at least three places in the world. Just over 5000 years ago, the Sumerians created what is considered the first script, cuneiform. Over 3000 years ago, Chinese characters appeared. Finally, just over 2000 years ago, the Maya created their glyphs in Mesoamerica. There are other examples, but these are the Big Three.

Only one can be traced from its origins in rock art to a script used today. That is Chinese. Cuneiform has passed through many forms and cultures, and somehow ended up in the Roman script I am writing with now. Although the Maya culture survives in Mesoamerica, knowledge of the script was lost.

Most contemporary writing systems are adaptations of

existing scripts, created to suit different languages. The Japanese imported Chinese characters, while the Greeks probably derived their alphabet from the Phoenicians. In Türkiye, the Ottoman Turks used an Arabic script, which was abandoned in favour of the Roman script in 1928. The Chinese adapted the Roman script in the 1950s, creating pinyin, but, as we shall see later, they resisted adopting the apparently simpler alphabetic script for written communication.[1]

The stories of the scripts that have been abandoned or adapted is a hugely complex array of ingenuity and is, unfortunately, beyond the scope of this book. Critically, I am not implying that cultures that did not invent writing were somehow inferior. The story of Mesoamerica will explain why that notion is so wrong.

After a quick look at what happened in Mesoamerica, I will compare two very different trajectories: that of cuneiform, which eventually ended up as an alphabet, and that of Chinese, which did not. Few scripts were ever used by other than a very small literate elite. Having a high literacy rate in a society is a very recent social goal.

Learning to write isn't about intellect, it's about need. That need arose when communities grew large. In most cases, 'writing' was simply a method by which land ownership, agricultural stock, trade and military records could be maintained, and, of course, by which lists of taxes to be paid were kept. It was bureaucracy that drove the invention of writing. The sheer volume of data, and the boring nature of it, meant that memory alone was no longer sufficient.

For many cultures there was simply no reason to develop a script, although the evidence is clear that all possessed the cognitive capacity to do so. Woollarawarre Bennelong, an Australian Wangal man, was captured by the British near Sydney in 1789. Seven years later, he was writing letters in English. Born around 1770, Sequoyah, a Cherokee silver-smith, created a script for his language. After a decade, he had a set of symbols representing every syllable. The Cherokee Nation went on to achieve a literacy rate exceeding that of the non-Indigenous settlers. Beginning in 1959, an illiterate sub-sistence farmer, Shong Lue Yang, created a script for dialects of the Hmong and Khmu languages in Laos.

Writing meant you no longer needed ritual performance to retain information. Knowledge could be transmitted not only across space but also across time, with neither the writer nor a messenger being present. But it also lost the performative element, and this was a much greater loss than anyone realised.

Writing in Mesoamerica: The Aztecs, Inca and Maya

Three civilisations existed around the same time in South and Central America: the Aztecs, who were literate, the Inca, who ruled without writing, and the Maya, who moved from a non-literate culture to develop their own script. All three ruled vast empires and all built their civilisations on cultures that had pre-dated them in the region. All created pyramid struc-tures in vast settings, which served, alongside other functions, as knowledge spaces for public and restricted performances.[2]

Narrative artwork was already a dominant feature of the Mesoamerican cultures before the time of the Maya, Aztecs and Inca. Around 2000 years ago, Teotihuacan, in what is now Mexico, was the largest city in the Americas. The incredible pyramids enclosed extraordinary artworks, including narrative murals such as the Tepantitla Mountain Stream (see Plate 10.1). The influence of Teotihuacan on later cultures, such as the Maya, is abundantly clear in their archaeology.

The Aztec cities, with their awe-inspiring architecture, art and gory customs, offer only a very short story. Around 1300 CE, the Aztecs settled in what is today Central Mexico. Their writing system was adapted from earlier Central American cultures and contained phonic elements. Drums accompanied their performances on pyramid-shaped performance spaces. About 200 years later, the Spanish conquistadors arrived and wiped out their entire culture. Before and during Spanish rule, the Aztecs produced highly illustrated books known as codices, along with a vast array of artworks.

There is an important lesson to be learnt for those who believe that writing indicates a 'superior' culture. Alongside the literate Aztecs and Maya, the Inca civilisation expanded to be the largest empire in pre-Columbian America, and did so without the wheel or draft animals and, most importantly for our story, without writing.

Like the Aztecs, the Inca had a brief reign, but it was extraordinary. From their heartland in the mountains of what is now Peru, their population expanded to be around 16 million, conquering a territory stretching over 4000 kilometres from what is now Ecuador, down the Andes mountains into Chile. Although best known for their spectacular city at Machu Picchu, their capital was Cusco. The Inca turned it into a massive knowledge space, which was documented in detail by the Spanish conquistadors before they destroyed it.

Over 40 pathways known as *ceques* (described in Chapter 4) radiated from the Coricancha temple, in the centre of Cusco.

It is not clear how much of the *ceques* could be physically walked and how much consisted of imagined sets of locations. Dotting the lengths of the *ceques* were 328 shrines known as *huacas*, some consisting of natural features and some human-made. Each *huaca* was encoded with information, much of it to do with pragmatic concerns such as irrigation, calendars and astronomy, but there were others to do with spiritual beliefs.

The heart of the city consisted of an architectural complex of platforms granting public and restricted performance spaces. Performances involved speech and dance accompanied by drums and shell trumpets.[3] Alongside the performance spaces and Inca *ceques*, memory experts used an extraordinary knotted cord device known as a *quipu* (also spelt *khipu*). *Quipus* stored numerical information, laws, censuses, tributes, trade, rituals and songs, genealogies and histories, and were used by runners to transmit messages throughout the huge empire.[4]

When the Spanish arrived in Cusco in 1532, the conquistadors destroyed thousands of *quipus*, and in doing so destroyed a culture.

Unlike the brief Aztec and Inca cultures, the first signs of the Maya date to about 4000 years ago. The culture spread, eventually covering a massive area that today includes southeastern Mexico, Guatemala, Belize and some of Honduras and El Salvador. The first signs of their gloriously artistic script emerged about 2300 years ago. The numerous and stunning Maya codices were highly illustrated and communicated all aspects of the culture through their elaborate hieroglyphs.

Spanish invaders arrived in Maya territory in 1521. In his zeal to convert the Maya to Christianity, Franciscan friar Diego de Landa ordered all their codices be burnt. In 1562 there was a grand incineration. Within a few generations, there was no one left who could read the symbols, despite the Maya culture somehow surviving. Four codices were hidden and eventually smuggled out; they provide us with a glimpse of the glorious libraries that were so tragically lost.

The Dresden Codex (see Plate 10.2) is twenty centimetres high. Folded accordion-style, the codex stretches out to 3.7 metres in length. The art is accompanied by text written in Maya hieroglyphs that describes local history, gods and goddesses. There are astronomical tables including eclipses and the movements of Venus, all associated with dates derived from the complex Maya calendar. Most of our understanding of the text comes from the Spanish conquistadors, who were not always reliable.[5]

There are many media for Maya art, much of which is three-dimensional.[6] For 1200 years, the Maya adorned stone pillars and narrative murals. The Murals of Bonampak, for example, tell the story of a successful battle and its aftermath. The victory celebrations included dancing, and trumpeters in an orchestra with drums, rattles and flutes. Speech scrolls unfold from the mouths of the singers, the undulation showing the note, intensity and nature of the sound.[7]

The Maya script was only decoded in the last 50 years. As a result, we can now read the Maya epic origin story, the *Popol Vuh*, almost certainly once written in Maya glyphs.[8] When the Spanish destroyed the codices, the Maya people retained the *Popol Vuh* in oral tradition until it could be written down again, this time in an alphabetic script.

By the time they were invaded by the Spanish early in the 1500s, the Maya and Aztec were under the control of powerful rulers who were notoriously violent. The similarly violent Spanish conquistadors destroyed these cultures, and along with them any ability for the Indigenous elite to read their own scripts.

Only a tiny portion of the Aztec or Maya could read. Very few of the Inca could read a *quipu*. Everyone else stored their knowledge in oral tradition. We cannot follow through what happened to the art, music, dancing and use of their architectural spaces as literacy expanded in the Americas. So we must leave them here and look at a script that tells a fuller story.

Falling in love with the Chinese script

Chinese is the only script that can be traced from the first marks to a script still used today. I realised that if I was to fully understand the long history of writing, from rock art to a modern script, I needed to learn how to read and speak Chinese. This seemed to be the hardest thing I could imagine learning, so would truly test my ideas on memory methods.

I soon came to the conclusion that this seemingly strange language and its unintelligible squiggles were so incomprehensible that they could not possibly function in the real world. Clearly, there was ample evidence to the contrary, so I persisted.

The massive initial hurdles were that this language had no alphabet, not a single word was familiar to me, and the inflection in your voice—which I continually failed to identify, let alone mimic—could change the meaning of a word completely. Then there were a whole swag of words that sounded exactly the same but had different meanings depending on context. I had to learn through a romanised phonetic version, despite no Chinese texts being written that way.

Then there was the script. I was told that I needed to recognise 3000 characters to read anything of significance.

I climbed that initial mountain and then, almost without realising it, Chinese all started to make sense. The fog cleared and I fell in love with everything about the language. To my absolute joy, the characters are no longer unintelligible squiggles but wonderfully intriguing and decipherable writing. My brain, which once had reeled, was now delighting in the language.

The earliest Chinese symbols become today's script

Gideon Shelach-Lavi, in *The Archaeology of Early China*, paints a picture of the Neolithic transition from small mobile cultures to sedentary farmers living in villages.[9] China, as a single country, did not yet exist. Despite the vast territory, there was the familiar pattern in the slow transition of egalitarian cultures to greater inequality, with elite burials appearing in the archaeology. Stone tools showed increasing sophistication, as did the artworks.

Cliff carvings at Damaidi in north-central China date back about 8000 years. Among the petroglyphs, there are 8453 different characters thought to depict social, hunting and agricultural scenes, along with the sun, moon, stars and gods.

As well as agriculture and the domestication of animals, there were elaborately decorated ceramics, jade carvings and bronze castings.[10] Incised tortoise shells and many flutes were found at the Neolithic settlement of Jiahu. A seven-hole flute made from the bone of a crane (*Grus japonensis*) could still be played, giving a range of an octave.

There were incised signs on tortoiseshells and bones found in graves dating to around 8000 years ago. There is a continuous record from these signs to today's script—an 8000-year sequence. I find that astounding.

Ancient texts record that the Chinese characters were invented by Cangjie, a legendary historian for the Yellow Emperor, who ruled around 5000 years ago. From then on, Chinese history is defined by its imperial dynasties.

Evidence of music has a long history in China. Of course, there is no evidence for the human voice, our main instrument, but some of the earliest drums known are from China, dating to over 7000 years ago. The Chinese made bronze bells, some of the earliest metal objects, created during the Shang Dynasty (3600 to 3046 years ago). Large, heavy sets of bells, each producing a different note and tone, were used extensively throughout Chinese history, becoming heavier and more complex over the centuries.[11]

Around 4000 years ago, the Neolithic culminated in the first recorded dynasty, the Xia, of which little is known. During the following dynasty, the Shang, characters were inscribed on bronze vessels, turtle shells and the famous scapulas referred to as 'oracle bones' (see Plate 10.3). The highly developed writing system involved about 5000 characters.

The oracle bones had been dug up by farmers in Henan Province and were sold as Chinese medicine until 1899, when scholars recognised their importance. About 4500 signs have been recorded from the oracle bones, of which a thousand or so have been identified. Many of the signs can be traced directly to a modern character.[12]

Around 3000 years ago, the Shang Dynasty gave way to

the Zhou (1046–256 BCE) and the script morphed into what is known as the Great Seal script.

As most of the population was illiterate, it is not surprising to see rock art still being produced. The 79 Huashan Mountain petroglyph sites display huge panels of closely packed symbols on steep cliff faces, some along the Zuojiang River and its tributory, the Mingjiang River (see Plate 10.4). These petroglyphs often start twenty metres above the ground or water surface, reaching 80 to 120 metres high. Their dates are still being debated, but it is generally believed that the rock art was started during the Warring States Period (475–221 BCE), around 2400 years ago, a time when the entire country was broken into about seven kingdoms at war with each other. Producing these petroglyph panels surely required a supreme effort.[13]

As has been seen on other continents, portable art was produced at the same time. The Chinese were superb bronze workers. Figure 10.1 shows a detail from a Warring States Period bronze vessel depicting ceremonial events with people playing bronze bells.

The Chinese were using the familiar combination of music and ceremony along with permanent and portable narrative art. As symbols became characters, these older technologies continued to serve for millennia.

Chinese folk performing arts

The script might have existed, but most of the population could not read, particularly in rural areas. They were 'illiterate' rather than 'non-literate', because there was a powerful literate elite. They still required their knowledge to be spread through performance.

FIGURE 10.1 Warring States Period bronze vessel, redrawn from Melinda Pap, 'Rituális étkezés az ókori Kínában', *Convivium*, Faculty of Humanities of Eötvös Loránd University, Budapest, 2012, p. 184. IMAGE: LYNNE KELLY.

In *Chinese Folk Performing Art* (2014), the authors describe an oral tradition dating back well over 2000 years.[14] This storytelling art form was accompanied by drums or stringed instruments. A single performer would take on many character roles as well as that of the narrator, using multiple voices, song, movement, rhythm and dramatic narration to tell the stories. The performance space was small, so the storyteller was very close to his audience. Sometimes there would be two performers, one acting as narrator and the other taking the roles of all the characters. The stories were very closely linked to folk dance performances.

The stories could be extremely long, often taking several months to tell. Even as recently as the 1920s, storytellers were immensely popular in China, and many were very famous.

Although some stories in more recent times were fictional, throughout history common themes included the social orthodoxy of Confucianism, morals and expectations, folk mythology, military heroes, legal cases and many aspects of the daily lives of the audience. Particularly common was the narration of historical events dating back to the early dynasties. Both children and adults learnt from the storytellers.

Figurines representing these performing storytellers are known from as early as the Warring States Period. During the Tang Dynasty (618–907 CE), the often obscene and vulgar content was shunned by the ruling classes, but, unsurprisingly, it was very popular among audiences.

Folk performers are recorded in paintings from the Song Dynasty (960–1279 CE), such as one of the best-known handscrolls, *Along the River During the Qingming Festival* (see Plate 10.5). This scroll is 25.5 centimetres high and 5.25 metres long. It depicts daily life in the capital of the Song Dynasty, Dongjing (present-day Kaifeng, in Henan Province). Included are 814 people, some of whom are folk art performers.

Folk performing was the most important medium for ordinary people to voice their opinions. The performers would literally sing the praises of great heroes who resisted brutal regimes. Satire enabled them to do so when outright protests would have been too risky.

Book publishing developed rapidly in the Ming Dynasty (1368–1644 CE), with many works of folk performing art being turned into written texts, while regional versions prevailed.

Folk stories around the time of the Revolution of 1911 used spoken words accompanied by a stringed instrument to tell of the revolutionary struggles faced by the army and people in the liberated areas. Even Chairman Mao Zedong came north to watch the performers.

Today in China, as is the case in all countries with high literacy rates, the need for music as a memory aid for critical knowledge has disappeared.

The script and the unification of China

As folk performers preserved knowledge for the majority, the few who were literate across the huge country, and who spoke many different dialects, were forced to use the same script. The characters were standardised for most of what we consider China today during the brief but brutal Qin Dynasty (221–207 BCE). The regime violently attempted to remove all traces of the Confucian tradition as they unified the country. The Qin introduced a less complex script known as 'small seal'.

Using their newly acquired centralised power, the Qin embarked on a swag of ambitious projects, including the Great Wall of China. They built a mausoleum the size of a city for the first Qin emperor with the incredible life-sized Terracotta Army: thousands of soldiers, horses and chariots. Fortunately, the horrific Qin Dynasty only lasted fifteen years, when civil wars again split the country until the Han Dynasty (206 BCE–220 CE).

'Standard' script is dated to the Han Dynasty, a time when the imperial civil service examination system was established. The script and examination system were to last for over

2000 years. Topics in these exams included classical texts on philosophy, law, astronomy, hydrology, mathematics, religion, medicine and military knowledge.

Through subsequent dynasties, with divisions and amalgamations across the vast landscape, writing served the bureaucracy while the vast proportion of the population remained illiterate. Standard script remained in use until the Communist Party rulers introduced the 'simplified' script in the 1950s (see Figure 10.2).

Mao Zedong and modernising the script

Chinese governments made many attempts to reach their dream of a common language. China was home to over 50 ethnicities and 120 dialects, many of which were mutually

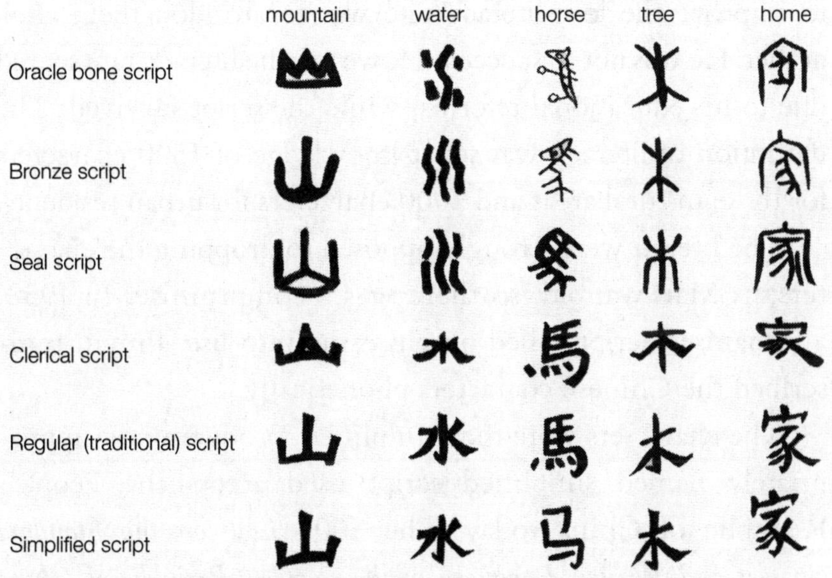

FIGURE 10.2 A comparison of the Shang oracle bone, bronze great seal, small seal, clerical (scribal), regular (traditional) and simplified scripts.
IMAGE: LYNNE KELLY.

unintelligible when spoken. Some local dialects had tens of millions of speakers. The literacy rate around 1900 was between 10 and 15 per cent.

The 1913 Conference on the Unification of Pronunciation tried to reconcile rival dialects and special interests. As the conference went on, verbal assaults escalated and became physical. One proponent of the northern dialect, offended by a misunderstood expression, pummelled one of the southerners with his fists and chased him from the assembly hall. After a month, the committee produced a solution that satisfied no one. And so the battles went on—a story told brilliantly by David Moser in *A Billion Voices* (2016).[15]

Mao Zedong proclaimed the establishment of the People's Republic of China on 1 October 1949. He believed that the Chinese characters were so difficult to learn that the only way to improve the low literacy rate was to abandon them altogether. He was not to succeed. However, the literacy rate soared due to his educational reforms, while the script survived. The definition of literacy was set at knowledge of 1500 characters for those in rural areas and 2000 characters for urban residents.

The literati were strongly opposed to dropping the characters, as Mao wanted, so there was a compromise. In 1958, a romanised script called pinyin came into use. Pinyin transcribed the Chinese characters phonetically.

The characters were then simplified, producing the appropriately named simplified script used across the People's Republic of China today. The 2001 *Law on the Standard Spoken and Written Language of the People's Republic of China* dictated that Mandarin, also referred to as Putonghua (普通话 or pǔ tōng huà), become standard for governance, education, mass communication and commerce, enforced through state-specified Putonghua examinations.

As a student of this script, may I take this opportunity to emphasise that the script is 'simplified', but not 'simple'? I find it very much not simple.

The old standard script, now referred to as 'traditional', is still used in Taiwan, Hong Kong and Macau.

Although it is technically feasible for pinyin to be used for all communications, it is not. One issue is that Chinese is a contextual language. There are a huge number of homonyms, words that sound the same but that have a different meaning depending on the context.

An extreme example is a famous poem written by Yuen Ren Chao in the 1930s, 'The Lion-Eating Poet in the Stone Den'. The Chinese character title is 施氏食獅史, which transcribes to *Shī shì shí shī shǐ* in pinyin. The poem was written in traditional script and pre-dates both simplified and pinyin. Critically, the poem is essentially unintelligible when said aloud or written in pinyin. But when you read the characters, it makes perfect sense.

In traditional Chinese characters:

《施氏食狮史》
石室诗士施氏，嗜狮，誓食十狮。
氏时时适市视狮。
十时，适十狮适市。
是时，适施氏适市。
氏视是十狮，恃矢势，使是十狮逝世。
氏拾是十狮尸，适石室。
石室湿，氏使侍拭石室。

石室拭，氏始试食是十狮。
食时，始识是十狮尸，实十石狮尸。
试释是事。

In pinyin, the marks above the letters are tones, more like intonation. They are not accents, as in French.

«Shī Shì shí shī shǐ»
Shíshì shīshì Shī Shì, shì shī, shì shí shí shī.
Shì shíshí shì shì shì shī.
Shí shí, shì shí shī shì shì.
Shì shí, shì Shī Shì shì shì.
Shì shì shì shí shī, shì shǐ shì, shǐ shì shí shī shìshì.
Shì shí shì shí shī shī, shì shíshì.
Shíshì shī, Shì shí shì shì shíshì.
Shíshì shì, Shì shǐ shì shí shì shí shī.
Shí shí, shǐ shí shì shí shī shī, shí shí shí shī shī.
Shì shì shì shì.

Translated into English:

'The Lion-Eating Poet in the Stone Den'
In a stone den was a poet called Shi Shi, who was a lion
    addict and had resolved to eat ten lions.
He often went to the market to look for lions.
At ten o'clock, ten lions had just arrived at the market.
At that time, Shi had just arrived at the market.
He saw those ten lions and, using his trusty arrows,
    caused the ten lions to die.
He brought the corpses of the ten lions to the stone den.

I am sure the literati had many other reasons why the script had to be retained, but I feel certain that part of their opposition was the culture of integration between the characters, calligraphy, art and communication (see Plate 10.6 for an example). The script is a significant part of a very long tradition of personal expression, intricately linked to the art forms over millennia. To abandon it not only made the artistic tradition unintelligible to future generations, but also meant a loss of self-expression.

The script is more than just a way to write words. The characters are much more than a medium for recording and passing on information. Calligraphy is a medium for creative expression, and the literate elite in China for the past few millennia, including the emperors, were expected to be skilled in painting, music and calligraphy.

Spanish artist Pablo Picasso said in 1956, 'Had I been born Chinese, I would have been a calligrapher, not a painter.'[16]

Even when writing in the script today, the obsession with getting the square exactly balanced reflects the origins in art. Components will be squeezed or stretched or reduced or expanded in order to fit neatly in a square. The more you get to know the characters, the more beautiful they look. From unidentifiable squiggles, they become collections of recognisable components combined into recognisable characters and then used as portions of recognisable words.

Words are composed of one, two, three and occasionally four characters. At first I could not work out why the Chinese put a gap between the characters that made up a word. If a word has three characters, why not squash them together and then leave a gap before the next word? Now I understand. That wouldn't look so perfect. Clarity is sacrificed for beauty.

A revealing moment came when I decided to write complex characters on cards to place around the house, in the hope that familiarity would develop. Writing with pens, even my calligraphy pens, I couldn't get the right effect. Chinese characters are traditionally written with a brush in a strict stroke order. Sticking to the rules, I wrote with a Chinese calligraphy brush. I was unexpectedly delighted with the result. Even at my beginner stage, it was almost a meditation.

I went to write the pinyin below with the same brush and it was dreadful. I resorted to a pen. I wonder how much the tools of the trade—a brush, not a pen—contributed to the resistance to change. Chinese characters are an art form every bit as much as a way of writing words.

Three millennia ago, during the Shang Dynasty, a significant proportion of the characters were pictographic. Many people believe this is still the case, but it is not so. Over 90 per cent of the characters in both traditional and simplified script are compound characters. That is, they include components that involve both meaning and phonetics. I struggle to see the 'pictures' in even those that are supposedly pictographic, except the most obvious, the ones everyone uses as examples, just as I have in Figure 10.3. In Figure 10.4, I have broken down the familiar greeting, *Ni hao*, to show the components.

How can a script work without an alphabet? In each character, one component acts as what is known as a 'radical'. Radicals can also appear as ordinary components within a character. Traditionally, there were 214 radicals, which served to indicate something about the meaning. Common radicals

FIGURE 10.3 Examples of characters in traditional and simplified Chinese scripts. IMAGE: LYNNE KELLY.

FIGURE 10.4 A word consisting of compound Chinese characters, *Ni hao*, broken down. IMAGE: LYNNE KELLY.

are mouth, speech and grass, not surprisingly, for things to do with the mouth, speech and plants, respectively. These are arranged in a radical table according to the number of strokes. This enables words in dictionaries to be recorded in order, despite the lack of an alphabet. The radicals also gave me a structure on which I could build my memory palace.

From China to Japan, the script is still an art form

The Japanese spoke a totally different language from the Chinese and had no script of their own. For millennia, it was expected that the Japanese literati could read Chinese. The Japanese therefore decided to base their writing system on the Chinese characters, which are referred to by the Japanese as *kanji*, an approximation of the Chinese name for their characters, *hanzi*.

The Japanese didn't include tones, but did invent a fairly small set of supplementary symbols known as *kana*. The *kana* show how a character is to be pronounced and how to transcribe native words. Technically, the language could be written in just *kana*, but the Chinese characters initially came with enormous prestige, and so have been retained.

The *kana* come in two flavours. *Hiragana* is used for normal writing. *Katakana* is used for words borrowed from other languages and for foreign names. It is claimed that a blind Japanese reader can read more easily than many with sight. This is because Japanese braille is written in *kana*, and not in *kanji*. The romanised version of Japanese is called *romanji*.[17]

Chinese script was also adopted by other East Asian countries, including present-day Vietnam and Korea. The shared script acted as a *scripta franca* to enable face-to-face interactions. Due to the mutually unintelligible languages, these 'conversations' were conducted in silence, through the medium of brush and ink. Known as *bǐtán* (筆談), which translates as 'brushtalk', these silent conversations were reported from the Sui Dynasty (581–618 CE) and continued into the 1900s.[18]

Unlike the Japanese, the Vietnamese and Koreans have since abandoned Chinese characters in favour of an alphabetic script.

From scrolls to manga

I believe the narrative art continuum can be traced back much further than 1000 years, from the earliest Chinese rock art via the scroll and Hokusai to manga.

Historically, both Chinese and Japanese have used art and script in exquisite long narrative scrolls. This format not only is superb in the original but also offers, I feel, huge scope for taking advantage of our biologically driven skill for using art in contemporary learning. I'll return to narrative scrolls in the next chapter.

I also wonder to what degree Japanese manga owes its popularity to the difficulty of the script. This highly graphic way of telling stories is hugely popular with adults as well as children. Picture books, like alphabet and number songs, have always been popular for teaching young children. Given our innate skill to associate art with knowledge and narrative, why shouldn't all ages enjoy all the enhancement offered by art to our stories?

In the book *One Thousand Years of Manga*, Brigitte Koyama-Richard traces contemporary manga comic formats back to the Chinese and Japanese narrative scrolls. Katsushika Hokusai is best known for his woodblock print series *Thirty-six Views of Mount Fuji*, including *The Great Wave off Kanagawa*. It was Hokusai who first coined the term 'manga', referring to his sets of drawings for training his students.[19] Hokusai's version of manga (see Figure 10.5) eventually morphed into the comic format so familiar today.

FIGURE 10.5 A sample of Hokusai's 'manga' from a woodblock-printed book dated 1834. IMAGE: THE METROPOLITAN MUSEUM, NEW YORK.

The *Ramayana*

Right across Asia, epics tell long stories, encoding vast amounts of practical and cultural information. They have been performed and painted from early oral times. One of the most famous of these is the *Ramayana* (see Plate 10.7). Even when literacy had taken hold, the *Ramayana* was still produced in a highly visual format.

Stories from the *Ramayana* are still performed and painted. Tourists regularly attend events such as the Kecak Dance, or

Ramayana Monkey Chant. This is a relatively recent Balinese telling of part of the *Ramayana* using dance, music and drama. The five acts of the performance involve a circle of dancers, often numbering over a hundred. Tourists from very different cultures engage with the unfamiliar story, yet the impact of the performance can be profound. One friend, Joan Newman, described it 'as feeling in a type of ecstatic state: heightened emotions, a feeling of being part of the performance, feeling a desire to move in a similar pattern'.

Contemporary artworks that dynamically reflect the performance are sold (see Plate 10.8).

The origins of the Roman script

This book is being written in English using the Latin (or Roman) alphabet. In a rash elimination of all complexity, I will oversimplify the sequence to say that first there was cuneiform, which may (or may not) have influenced Egyptian hieroglyphs, which inspired the Phoenician script, which gave rise to the Greek alphabet, which led to the Roman alphabet and to all modern European scripts.

I would so love to exclaim that the origins of the script I use lay in recording our human thirst for knowledge, to preserve forever the philosophy and thoughts of the wise, the history of our species and our insight into the natural world. But I can't. The sequence probably began with accountancy.

It all starts with the Sumerians, who built their city-states around the delta of the Tigris and Euphrates rivers, in present-day Iraq. From about 9000 years ago, early farmers used simple images to label products for trade. As cities grew, manufacturing led to more specialisation and they needed more complex tokens. A blunted reed was employed to make the impressions, leading to 'wedge-shaped', or cuneiform, writing.

The cuneiform script spread and was used as a writing system for about fifteen languages over 3000 years. It was, at times, the script of the Babylonian, Syrian and Persian empires.[20]

Epics were written down early in the development of writing. Cuneiform is no different. The earliest cuneiform clay tablets telling the *Epic of Gilgamesh* date to about 4000 years ago.

As with Chinese, the aesthetic form of the characters seems to have been of huge significance. Although the wedges could have been made with the stylus at all angles, only a limited number of angles were used. The writing provided stunning works of art.[21] Plate 10.9 provides just such an example. Not surprisingly, given the importance of music and art in early knowledge systems, musical notation during the Old Babylonian Period, nearly 4000 years ago, was recorded in cuneiform.[22]

The Egyptians did it differently

Where Egyptian hieroglyphs came from is still a bit of a mystery. Some believe the interactions between the Sumerians and the Egyptians led to the latter copying cuneiform but doing it their own way. Others think the Egyptians invented writing themselves. I don't know. I'll have to wait for the experts to work it out.

Of course, that won't stop me speculating. There are some early pictograms that closely resemble certain hieroglyphs. These pictograms included representations of land, villages, mountains, stars, moon, Earth and totems of tribes. They appear on pottery, weapons and tools in pre-dynastic Egypt. I suspect the Egyptians loved their pictographs, saw what was possible with cuneiform and adapted what they already had in a burst of inventiveness. By about 5000 years ago, the Egyptians were using a full set of hieroglyphs.

But hieroglyphs weren't the script that Egyptians used for standard documents. Almost from the time of the first hieroglyphs, two cursive scripts appeared, hieratic and (later) demotic. Demotic Egyptian appeared on the Rosetta Stone along with Greek and hieroglyphic versions of the same text, enabling the deciphering of the hieroglyphs.[23]

Hieroglyphs are wonderfully artistic, superbly visual. They could be drawn facing either direction and made symmetrical to fit aesthetically with the objects on which they were inscribed. Is this part of the reason why they survived for many centuries after there was widespread use of alphabets?

The Phoenicians used Egyptian hieroglyphs as inspiration, but replaced them with symbols for sounds, giving us the first alphabet. From about 3000 years ago, their brilliant idea took hold and spread. Their alphabet consisted of 22 letters, all consonants. The Phoenicians traded widely around the Mediterranean, inspiring many other cultures to develop alphabets. It is not fully understood how the Roman script appeared and travelled to the modern world, but it was the Greeks who added symbols for vowel sounds.[24]

The artistic aspect of writing was lost as the script grew to represent the sound of our speech alone. The skills offered by the language gene, FOXP2, dominated our learning and the way we recorded knowledge from this time on. Art still represented narrative, especially for the non-literate, and was used to make written records more memorable. But for the educated elite, it was all about writing in an alphabet that replicated the sound of our speech.

Epics, originating as oral tradition, were written down. As the *Popol Vuh* was written in Maya glyphs and the *Epic of Gilgamesh* in cuneiform, by the time of Homer epics were being written down in Greek.

The ancient Greeks write down their stories

It is believed that Homer was born around 800 BCE, but there is doubt as to whether he was a real person. The name may have applied to an amalgamation of the bards of his era. Something else I'll have to wait for the experts to decide. What is not in doubt is that Homer represented an epic tradition. He was essentially the encyclopaedia for his society, using narrative to relay issues of governance, education, social expectations, law, history, religion, customs, geography, science, technology and knowledge of faraway places.

The bardic song-poetry was composed in a way that made it easier to memorise, but also that was entertaining so the audience would want to listen. There is a repeated beat to the performance. The story is sequenced, with each scene set in a specified location, telling the actions of vivid characters. The journey itself acts as a memory palace, as the players move from location to location.[25] This is an ancient form of the method of *loci*, as described in Chapter 4.

Homeric bards could recite the entire *Iliad* from memory— nearly 16,000 lines. It is estimated that it would have taken at least four long evening sessions to recite in full. *The Odyssey* was nearly as long.

A few hundred years later, another star of ancient Greece was railing against writing. Born in 469 BCE, Socrates did not write. His words were recorded by his pupil Plato. In *Phaedrus*, Plato presents Socrates quoting a conversation between the Egyptian credited with the invention of writing, Thoth, and the god Thamus, the king of Egypt. Thoth said that the invention of letters would make the Egyptians wiser and improve their memories.

Thamus disagreed: 'For this invention will produce forgetfulness in the minds of those who learn to use it, because they will not practise their memory. Their trust in writing, produced by external characters which are no part of themselves, will discourage the use of their own memory within them.'[26]

Socrates was so right! That doesn't stop us grabbing back lost skills, and keeping writing. But I digress.

Homer's mnemonically valuable rhythms became music education with a different emphasis. From Plato's time on, music education was justified in terms of its ability to influence people to be more effective citizens. Into the Middle Ages, a period of Church dominance over every aspect of life, musical knowledge was seen as an essential part of the education of a responsible adult, one who followed the teachings of the Church. Charlemagne, crowned Holy Roman Emperor in 800, decreed that schools for boys should include 'the Psalms, the notation, the chant, and arithmetic and grammar'. It wasn't until around the middle of the 20th century that educationalists no longer felt that music served society's needs and goals. From that time on, musical education has promoted aesthetics as its fundamental goal.[27]

In late republican Rome, the memory methods based on spatial skills were still taught. Born in 106 BCE, Marcus Tullius Cicero was considered one of Rome's greatest orators and lawyers, hugely admired for his phenomenal memory. Writing about memory in *De oratore*, Cicero did not bother to explain the method of *loci*, claiming that to do so would be tedious as it was already familiar to his readers. Associating knowledge with place was part of everyday education and life. In the next chapter, I will argue that it still should be—we are wasting a skill hard-wired into our cognition.

Many Greek performances involved narrative, dance and music. They were also represented in art forms, such as on the famously beautiful Greek vases. The geometric designs on the vases of Athens from 900 to 700 BCE were gradually replaced by images of the narrative progressing around the vessel, such as the portion shown in Plate 10.10. Over 1000 years or so, the images came to represent a single static scene.[28]

I have spent many pleasant hours in museums that lay out their Greek display chronologically, such as the Metropolitan Museum in New York and the British Museum in London. I'm generalising, I admit, but it is clear that the art changes from mnemonic to narrative to decorative as literacy takes hold.

As literacy dominated the elite classes, oral technologies still served the illiterate members of the community. Between 1933 and 1935, Milman Parry and Albert Bates Lord likened the 'singers of tales' in the Balkans to the ancient Greek bards, because their performance methods so strongly resembled the Homeric techniques: they told their traditional narratives using a stock of verbal formulas recited with rhythm.

Lord described these improvised poetic performances as being an oral tradition.[29]

However, there is a significant difference between the Balkan bards and the Elders in oral cultures. The Balkan bards were illiterate performers in a literate culture, continuing the oral tradition for a predominantly illiterate audience. This is significantly different from purely oral cultures, where those with power are oral themselves, not literate. The 'singers of tales' were using the techniques of oral tradition, but much of the purpose and content had now been taken over by writing.

In today's highly literate societies, there are still traces of the ancient bardic traditions. American jazz critic and music historian Ted Gioia compares the bards of Parry and Lord's analysis with contemporary jazz performers.[30]

Gioia talks about the way the bardic singers depended on using set phrases over and over to satisfy the demands of creative improvisation under the pressure of a live performance. He sees exactly the same process when a jazz soloist demonstrates extraordinary mastery to fit set phrases into different contexts. The result sounds, to those of us who are not jazz performers, like spontaneous improvisation, but Gioia says that the skill often involves a substantial repetition while cleverly juxtaposing formulae. This is a concept to which I will return in the next chapter.

Medieval art was about knowledge and memory

Moving into the Middle Ages in Europe, education was dominated by the Catholic Church. Music became a tool of religion, reinforcing spiritual and moral teachings. Power was firmly in the hands of the literate.[31]

From early in the Middle Ages, students were expected to memorise huge portions of the Bible, along with important speeches, both contemporary and ancient. They needed to study law, history and the natural world. There may well have been books, but they were handwritten and hugely expensive, so students relied on their wax tablets, which had to be constantly erased. They relied on memory and their spatial skills by using memory palaces.

The purpose of art as a record of information was reduced from prehistoric times, but huge panels of mnemonic images were still around in wondrous, ornate, grotesque, vivid, elaborate stained-glass windows and dominating sculptures. Images in panels on church walls were laid out in grids, the physical positions in the painting making them more memorable.

All genres of art were used in various ways to make the text in books more memorable. The highly decorated medieval manuscripts are some of the most glorious artworks ever produced, but their primary purpose was to convey knowledge (see Plate 10.11).

There are lessons for us about making written text memorable on every magnificent page. The medieval scribes added all sorts of ornamentations around the text, such as images of grotesque and fanciful beasts, strange figures displaying gross ugliness and extraordinary beauty while often enacting vulgar behaviours.

Highly decorated 'glossed books' were laid out specifically for memorising. The glosses were essentially notes added by various readers in between lines of the original text. Often quirky illustrations of animals and people enacting violent, humorous or titillating stories, known as drolleries, were added around the text to make each page look different.

Depictions of animals abound in the highly illustrated medieval bestiaries. Some critters were clearly real, others far from it, yet the distinction is not clear. The beasts were employed to represent moral stories, acting as a mnemonic for those who could not read the text and enhancement for those who could. The love of bestiaries continued well into the Renaissance. I use my own version of bestiaries to recall anything with words, from names to vocabulary for foreign languages.[32]

These days we too often try to learn from page after page of typed text with white borders and not an illustration to be seen. Sometimes we highlight passages, but that is still repetitive and dull. I am well aware of the irony of that statement and the format of the book you are now reading.

From the Middle Ages into the Renaissance

As the Middle Ages melted into the Renaissance, printing helped increase literacy. A plethora of short books known as memory treatises, or *ars memoriae*, hit the market. These handbooks emphasised the use of places and images to memorise all manner of information. Advice for creating memory palaces, which remains useful today, was offered in the 15th century by Jacobus Publicius in his extremely popular manual on rhetoric, *Ars memorativa* (1482):

> Similarity among the locations should be avoided more than death . . . For these reasons, we should be able to avoid this by [adding] color to the structure and height to the figure, and by [using] diverse material. Or if we do not do so by

means of places chosen and arranged with art, at least our places should be fashioned with variety, using stones, streets, mounds, altars, monuments, bireme boats, inclined bridges, stars, and islands.[33]

Well into the Renaissance, the knowledge gene NF1 was still being used to far greater effect than it is today. The books were no longer in colour, and eventually many were produced without illustrations and without concern for memory at all. Music gradually changed its role as well. Musical notation has remained all symbols. There is no alphabet and no phonetics. Yet it can be read in such a way that a skilled player can reproduce, note for note, what the composer intended. And it can be read by readers of any language.

Renaissance hymns and songs could include as much repetition as they liked. It was not only nursery rhymes, ballads and popular songs that love choruses and repetition. Classical music, from the Renaissance on, is endlessly annotated with repeat marks and variations on themes. Slowly, music's purpose evolved. As writing dominated learning, music became much more a form of entertainment. Big themes have become rarer, and most pop music is what John Lennon described Paul McCartney as writing: 'silly love songs'.

Analysing my own music collection, I was surprised to find that well over 90 per cent was 'silly love songs'. This includes popular songs, opera and most things in between. I have no objection to silly love songs. They have formed the soundtrack of my life and will continue to do so. But we are still hardwired to use music as a memory aid for critical knowledge. We just don't do it anymore.

Music has gradually left the encircling community and climbed up onto the stage.

But hieroglyphs are making a comeback. We use a multitude of symbols that academics describe as being modern hieroglyphs.[34] Street signs, travel icons and mathematics all use symbols rather than alphabetic words. As you read non-phonetic symbols such as +, &, $ and 5, you will be saying words to yourself. A vast array of emojis has also taken advantage of our cognitive response to meaningful art.

We have expunged music, art and our natural connection to physical locations from our learning toolkit because literacy appeared to serve all our knowledge needs. It doesn't. Literacy has brought us so much, but it is also a two-edged sword when we consider the longer-term story of its impact on human culture. Literacy has brought us communication systems that can extend to so many more people than was even imaginable before. But we need to listen to those who think differently. We need to learn from the performers and the musicians and the artists and the storytellers.

We can regain what we have lost and adapt those lessons to the world we live in now. The next chapter will explore how to do just that.

CHAPTER 11

Putting our knowledge gene to work

We have a choice. We can keep lauding those whose thinking fits within the mindset that has dictated our lives for centuries. We can go on rewarding only those who sat quietly in classrooms, wrote their notes neatly and passed their written exams with flair. In that way, we can go on making the same mistakes and lacking the solutions to the problems facing the world today. Or we can do so much better.

I sat wonderfully comfortably in the typical education world. I loved school and university and teaching and learning. I loved sitting still and writing my neat notes. I passed my exams. I loved feeling that my orderly world was rather more important than that other world of art and music, and that I was more important than those people who didn't behave in our quiet, orderly classes.

People like me have a role, but it is not the only role. We all lose so much by not listening to different voices and not incorporating their ways of thinking into the decisions that affect the entire population, not just those of us who sat calmly at our school desks. Every leader would benefit from having their thinking knocked around by sitting down with those outside the halls of power.

I believe that the only way to find creative solutions is to take advantage of human diversity, just like those early human populations, who adapted to almost every biome on the planet. We need diverse thinking, and we need all those diverse thinkers sitting at the same table.

There are a lot of voices other than mine in this chapter. For the last two years, I have been meeting with those I call my Advisers. We've met individually and in groups crowded around a table laden with food. Between them, they have vast experience in music, art, performance, neurodivergence, education, parenting and the challenges of ageing. The Advisers have enabled me to think from a whole lot of different perspectives way beyond my experience. Unfortunately, I can only represent a fraction of their insights here, and even less of their unique personalities and the cross-pollination they brought to this project.

They say we should learn from the past. That past does not stop at 100 or even 1000 years ago. It goes back tens of thousands of years. We have stopped using so much of our innate potential. We have stopped taking advantage of the diversity in human populations. We have stopped working as a community to benefit from art, music, our sense of place and our vast cognitive potential. We have stopped taking advantage of what made us human in the first place.

For the sake of the world, the time has come to listen. If only we had listened long ago, rather than silencing different voices, we would be far better off.

Learning from First Nations voices

Frida Jennings is a twelve-year-old Canadian from Toronto. I had never heard of Orange Shirt Day, so she explained by telling me the story of a young girl from the Stswecem'c Xgat'tem (Canoe Creek/Dog Creek) First Nation in British Columbia, and her first day at the St Joseph Mission Residential School:

There was a lady called Phyllis Webstad and she was being sent to a residential school for First Nations people. It was 1973 and she was six years old. Her grandma bought her a brand-new orange shirt for school, which for Phyllis was a very special thing.

Phyllis thought that going to school was going to be a cool, fun experience and she was going to make lots of friends and learn new things. When she got there, they took her orange shirt away and made her wear a uniform. It was the kind of school where they wanted to make the kids forget their First Nations heritage, culture and language.

They were fed very poorly and didn't get a lot of clothing and they didn't have medicine so if anyone got sick, they just had to brave through it and that's one of the reasons a lot of the kids who went to residential schools died. They died of medical experiments and they were treated like test subjects. And they died trying to run away back to their families. Even now they are still finding the children's unmarked graves. Hundreds and thousands of these children that have just been taken from their homes.

The last residential school only closed down 27 years ago. The Phyllis Webstad story is one of the most known ones and that's why every September 30 we wear an orange shirt to remember that story and commemorate what has happened in a bad way.

Like it says on the shirt: every child matters. Every single child from everywhere, no matter their heritage, no matter their culture, no matter how they grew up, they still matter. You can't just force someone to take away their heritage, because that's how they grew up.

If only we had listened and learnt from the First Nations cultures in Canada and Australia and the United States and everywhere in the world, then we would understand so much more about different ways of knowing. And, even more importantly, the atrocities experienced by First Nations peoples may have been avoided.

First Nations knowledge technologies can be transferred to contemporary learning without misappropriating culture. You just need to be careful to respect restricted knowledge.

David Kanosh was able to transfer the traditional skills his grandparents had taught him, as described in Chapter 4, to a conventional classroom:

As I was about to start attendance at Sitka High School in the mid-1980s, my friend Rick Sunde (non-Tlingit) had been curious about my memory devices and asked if it was like the memory palace technique. Rick then demonstrated his skills with the memory palace technique. I had the impression that everybody else pretty much had the idea of memory skills.

I had finished the majority of required classes quite early. This opened the options of taking elective classes so for this I enrolled in the language classes to learn French, Spanish and Japanese all during the same school year.

I made memory boards for each, started weaving strange stories using all the vocabularies, singing and dancing the vocabulary lists located at the back of each textbook.

I paced myself but at the end of two weeks, I had memorised completely the vocabulary lists contained in the back of each textbook. The rest of the school year was spent listening, speaking and writing the necessary assignments.

I had completed each class with high marks and, oddly, to the amazement of the teachers. I wasn't sure what they were mystified about. I repeated the process for the next couple of years for these classes.

Accusations sometimes came up that I was somehow cheating. They never proved it. They never asked how I learned.

When Rick had demonstrated his skills with the memory palace, I had just assumed that everybody had access to this information and was using it. I never really thought about it.

Only a month ago did a former teacher ask how I did it.

My colleague Tyson Yunkaporta wrote:

Indigenous worlds of knowledge require that most of a person's workload consists of the maintenance of land and dense kinship relations that provide the infrastructure of memory. Obviously it would be impossible to build an entire Indigenous memory framework into an industrial or even post-industrial lifestyle that demands individualism, fractured kin affiliation and limited access to landscape in life and work. However, it is possible to reinstate some of our human evolutionary affordances to improve on the current disconnected practices associated with knowledge and memory. A possible path through that process might be as follows:

1. The knower creates symbolic maps of the routes they travel regularly in the places they live/have lived.
2. The knower translates the information they wish to remember into stories and encodes these stories onto their maps.
3. The knower creates phrases utilising rhyme, repetition, alliteration and taboo language/content to enhance memory at different stages of the story map.
4. The knower creates objects and images (or both, inscribing images on the objects) that reference maps or parts of maps as memory aids.
5. The knower creates these things in collaboration with others who are learning the same information, forming warm relationships in which the knowers keep the memories together.

However, the largest limitation is in the way landscapes are constantly altered by civilisations. You might find that the tree where you stored the Greek alphabet is cut down, or the street where you kept all your integers is demolished and replaced with a car park. You might find that you have to move to another state to find employment. When this happens, I find that people begin to understand the experience of dispossession for Indigenous people, and learn what it is to keep alive the knowledge of landscapes and ways of living that now often exist only in memory. It can be painful, but it still works better than most other memorisation techniques.

Although language is hugely valued, the role of art and music, memory methods and human connection have been sidelined in contemporary education. These should be at the heart of the curriculum, serving the whole curriculum, as they are in Indigenous cultures.

Recent research on dementia indicates the depth of musical memory and recognition of place, which are retained long after many other cognitive domains are gone. With an appreciation of our ancient knowledge gene, this is no surprise.

It is easy to sit still and write notes. It is so much more effective to supercharge learning with music, art, story and memory palaces. We all have a massive untapped potential available to us, from the time we are toddlers to when we're a long way through our life's journey. So how exactly do we do that?

The time has come to enhance our learning by engaging with all that our knowledge gene offers. This chapter is all about practical methods to do so at school, university and throughout a long life.

Starting with the young

From very early in prehistory, humans have made sculptures representing themselves and other animal forms.

Paul Allen describes himself as an artist-teacher who applies the qualities of his creative practice to teach artistically. He teaches in primary schools and adult classes. In 2017, we worked together on a project exploring ways to integrate art practices across the curriculum with primary school students. We decided to help the students tell stories about their learning by using figurines, as is common in Indigenous cultures.

We had trouble deciding what to call the sculpted characters. They were modelled on the *kachina* dolls of the North American Pueblo (discussed in Chapter 8), but representing mythological characters in art is universal among oral cultures. We wanted to be respectful, so any term that indicated a particular Indigenous culture was rejected.

We also wanted Paul's students to have a sense of fun. Paul suggested the name 'rapscallions', and it stuck. It's been used by many others since.

As part of Paul's art curriculum, the students created their rapscallions from a stick frame, plaster and decorations. We then used them to tell stories for science and mathematics. Creating characters from natural materials is something no other species can do. But they don't have our knowledge gene.

Paul has continued to develop the way he uses these characters in schools:

I now use papier-mâché and shredded paper to create the form because it's a lot lighter, but it's still basically the same process. The kids start with sticks, usually nice shaped twisted sticks; there are plenty of bent gum tree sticks in the school grounds. As they form these sculptures, ultimately, even if they may not have deliberately started with that idea in mind, they will see the gun or they will see the horse head. The capacity to look at a stick and see it as a horse's head or a gun or a figure—that is innate.

You can't play with a picture of an animal the way you can with a sculpture that you can actually hold in your hand.

That seems to be the difference with three-dimensional sculpture. They want to play with the figures so much that I have to tell them to put them down because they're going to break them before we've even finished making them.

One young student, let's call her Mary, made a cross-armed devil which she called Paul. He was her devil's advocate. While she was writing a persuasive text, he was the guy that she would use to be the other debater in her mind. Naming it after me really fitted with her personality; she was quite the contrarian. It almost made it valid because it was no longer her having the idea. It externalised thought like a puppet or another figure. She could argue with this figure and then just write down that argument versus having to do it in an abstracted way. She was convincing him, beating him versus something weird like writing to some abstracted concept.

The rapscallions seem to give the kids permission to be wrong about something. It was the rapscallion who was making a mistake, so this externalised object gave them the freedom because it wasn't about them. This was especially valuable for a child who had self-confidence issues from a traumatised background. That freedom and that disassociation allowed his intelligence to show. His learning difficulties were a psychological difficulty of 'oh, I can't do this'. He couldn't remember the times tables because he had all this stuff in his head about how he couldn't do it, but when his rapscallion learnt, the rapscallion remembered that six sevens are 42.

I am convinced that every school should set up a permanent memory palace. This need not involve any physical changes in the school. All that is needed is a map that numbers locations throughout corridors and the school grounds.

With teacher and class using the same palace, communal learning can happen, with chatter during encoding offering all sorts of creative possibilities. Multiple layers of data can be included at any location, so the same palace can be used by different classes. Poles can be painted and corridors decorated with knowledge. Performances can take place at all the memory sites, which will then take on the feel of being sacred, special and worth protecting. You don't need a stage. Song, dance, story and art will permeate the school as it becomes a massive knowledge space.

During our joint project in 2017, Paul and I established a memory palace at Malmsbury Primary School in Victoria, where Paul teaches. We talked in November 2023 and he told me how it was going:

Last Friday we were doing the invention of the telephone on the memory trail. You might remember that it was the little air vent from the heater. As the children walked under it, I asked the children all to pretend they were ringing a bell to associate with Alexander Graham Bell, and then answer the bell. The kids delighted in it. They were screaming and laughing and having a ball.

First, they pretended there was a bell to ring on the phone to answer. They said silly stuff which didn't have any relationship to the knowledge. They knew what mattered was that location was for the invention of the telephone.

The children are almost like improv actors because they make things up given only the smallest number of directions. I believe it's innate; kids are just playing. It's just what they do.

When we work on the memory trail, I have noticed the kids had to be of a certain age, around the seven to eight year mark, before they could create their own metaphorical association. Before that, the metaphor needed to be given to them. One little boy, I'll call him Quinn, was doing the different dog breeds. He was initially having struggles, as even adults do, until he got the hang of it. Eventually, he was coming up with his own associations. For bulldogs, he was running around like a bull. For golden retrievers he ran and fetched gold. He was in Grade Two, whereas when we worked with the Preps and the Grade Ones, they couldn't do that.

I have witnessed underachieving students thrive when given a memory palace to encode and recall. This was particularly evident for those diagnosed with dyslexia, whose spatial skills put them at an advantage over most others when remembering anything associated with physical spaces. Being familiar with the language and topic in advance was invaluable when later faced with the written words.

Similarly, the physicality of memory palaces engaged those whose worlds are best served when active—such as those with ADHD. I have yet to meet a student who did not find a memory palace effective.

A memory palace costs a school nothing—they already have one in their buildings and grounds. And education should not stop when you leave the hallowed halls of learning. Every home, and the streets around it, offer a palace to anyone who wants one.[1]

Memory palaces for every age

I have been using memory palaces for over a decade now, and am constantly astounded by how much I can recall. More importantly, I am astounded by the questions I ask because of the patterns I see. Instead of delving down to a fact, as you do when you use a search engine, I have a big picture floating around in my head and want to know more.

Each morning, I walk one of my memory palaces, learning Chinese or French vocabulary, reviewing 1000 digits of pi, adding more details to my History Walk or my Countries palace. Exercise is essential, but I am a reluctant walker. When I learn as well, the time flies. I am exercising body and brain.

Paul Allen teaches adults as well as small children. He has been using memory palaces with adults for a few years now, for many different topics. He always gets the surprised reaction I had when I first tried the technique. And no one is too old, as he explains:

I was teaching an adult class on the twenty elements and principles of visual art. I used a memory palace with five items per wall in the drawing and painting studio. The students were coming up with their own versions of things: form, lines, shapes, space, colour, textures, balance, time . . . After a break, I quizzed them and, of course, they all got it. They could do it backwards and just jump to the seventeenth item, say, and shuffle it around very, very quickly. This is really useful when analysing their artwork.

At the end of that day, a chap came up to me. He was an older man and he said he didn't know how I had done it. He had left school at fourteen because he was always the dunce because he could never remember anything. He said, 'That's the most amazing lesson I've ever had, because I've never been able to remember anything before.' He was in his sixties.

When we did the actual sculpture, we made use of those elements and principles, some more than others. The impact of that learning over six months of the class just meant that when this chap was considering his work, those elements were on his mind—for example, texture in the clay sculpture casting. He was very aware of the principle because he had the information in his head. He could then apply it. He was applying them in his actual work, in the context of the knowledge.

Rhythm, rhyme, repeat, remember

The primary school students at Paul's school had learnt the definition of the word 'force' in science: a push or a pull. The music teacher had encoded the definition into a song. They were all capable of correctly defining force.

But does that knowledge stick? Paul told me: 'In 2022, the kids who had learnt the push and pull song in Grade Two were waiting outside a classroom. They were all just singing the push and pull song, just out of the blue, four years after they had learnt it.'

I've heard older students repeat, ad infinitum, a chemistry chant about acids and bases mixing to give salt and water. And, just like the much younger students, as they beefed up their chanting with familiarity and movement, they were laughing.

One of the best-known versions of curriculum-by-song is when Tom Lehrer sings the chemical elements. Great as the song is, the priorities are different from those in the classroom. Lehrer chose the order of the elements according to rhyme and rhythm. Far better for education would be to sing them in the order of the periodic table. The problem with a song of the elements is that you can't layer further information. This is where you turn it into a songline. Singing the elements in order can then link them to a numbered memory palace. Once an element is associated with a location, anything you want to add about the element can be layered at the location.

Learning for the long term requires repetition. It is worth repeating that we will tolerate repetition in music much more than in speech. I have been watching videos of concerts of performers whose music I enjoy immensely. The first time they sit on the stool and tell me a bit about themselves, I find it really interesting. The second time, maybe not so much. By the third time I am irritated and just want them to get to the songs I have heard so many times before.

Movement is critical, too. The research on this is unequivo- cal. Humans universally feel the urge to move to music. While visual art is manifest across space, music is manifest across time. Being carried away by the music through movement or through reverie is usually because of the temporal qualities of the music. The pulse, tempo and rhythmic patterns are what we respond to. This is why music literally *moves* us.[2]

Using familiar songs to encode useful information works wonderfully well. I might be out of tune when I sing my French vocabulary songs, but I remember the words. I knew that warbling refrains was great for memory, but I had no idea just how much more it offered. For schools, the evidence is robust that both children and adults benefit hugely from getting together with others and making music. Singing, group drumming or dancing—it doesn't matter what. Just make sounds and move with rhythm.[3]

Implementing music within learning could be anything from just speaking with rhythm, adding the rhythm through an instrument, adding rhyme, making a chant or going for a fully-fledged song. Some people are comfortable with singing, some less so, but there is no culture where people in a group won't chant or sing in time with each other.

Mary Thorpe is a performing arts coordinator mostly working at primary school level. She gave me her thoughts on this matter:

At some point everyone must be able to play their part accurately and each kid has to keep the beat/play in time (preferably play in tune and preferably sing in tune, too). If kids can't feel the beat, or time, we just keep sticking with it until they do. They all get it eventually. They may have to watch another kid to be prompted, they may not listen, feel and hear as well as others, but the song/musical piece needs everyone to play TOGETHER or it just falls apart or never gets going. If it is not together, then it is not music, it is noise.

Students have more opportunity to bond as a group when they sing together in their classrooms. There is no need for scrutiny or performance; it is enough to do it for pleasure and wellbeing. Simple percussion could be available in every primary school classroom—clap sticks, shakers, little drums.

Students find it easier to hear nuances in song than in speech because the rhythm is slower, making it an effective tool for the teaching and learning of language.[4]

I have no doubt that, had I asked my physics class to sing, a few would have been delighted and most would have been horrified. But had I asked them to chant, much as they do at a sports match, I would have had the entire class in tune. I am not suggesting turning the curriculum into some grand opera, but difficult words, phrases and important concepts can be turned into a chant.

In the classroom and throughout life, I want people to get rhythm, add rhyme, then repeat and repeat so that they remember.

But there's even more. Music also has a massive social and emotional impact. As discussed in Chapter 7, the prehistory of music encourages us to think of it as a social bonding device as much as a memory one.

Dopamine is your free, internally generated wonder drug with no side effects. It is our reward hormone. It not only invokes a sense of pleasure but also serves as a motivator to act. Making music increases our levels of dopamine and related hormones. And, as a bonus, it reduces our production of cortisol, the stress hormone.[5]

Hilary Blackshaw is a music educator with secondary students and a violinist who performs regularly in various genres. She is particularly concerned about the stress levels she sees in students. Hilary is convinced that music can play a significant role in alleviating them, as she explained to me:

> When we focus on the process of making music, rather than the outcome or end product, connection is brought right to the forefront. To engage in the process of making and playing music in a meaningful way, we need to be able to connect with what we are playing, connect with those we are playing with and also for those we might be playing for. Music that is process driven, rather than fixed outcome driven, will often rely on an external stimulus, story or spark of inspiration which will foster that connection to engage deeply with the music. This creates a stronger sense of agency for everyone involved.
>
> Within an educational context, a group of young players can be brought together to play music, often not necessarily having strong, if any, social connections initially. There is often caution, hesitancy and a reticence during the early engagement phase. But as young players are supported to explore a narrative or theme through the music, and therefore bring their own experiences, ideas or stories to make meaning and form connections to the music, a subtle shift begins to occur. They come together as a group, socially, creatively and musically. They connect to each other and the music through ideas, communication and problem solving.

But what is most interesting, they are often able to overcome some quite substantial anxieties, challenges and fears through feeling connected to something bigger than themselves; the group, the dopamine hit of playing music together, the sense of achievement and a connection which will often create memories lasting for many years, if not a lifetime.

By allowing the students to follow the process, taking time to explore themes, ideas, communicate, and tap into the deeper level creative and critical thinking parts of the brain and nervous system, something quite magical happens. It's a process which draws in everyone. The learning really is about the journey, not the destination.

Mary Thorpe experienced the impact of music on young students isolated at home during Covid lockdowns in 2022:

During lockdown I turned up each week wearing a costume that fitted the theme of the unit of work and I asked students to dress up as well. For example, for Grade 1–2 we did a unit of work on *Carnival of the Animals*. We pretended to be zoologists, wearing costumes. We discussed the elements of the music and how the music matched each animal; the beat, rhythm, pitch, instrumentation and dynamics. Each week we explored a different animal and its connection to the *Carnival of the Animals*, adding in drama or dance.

As students were waiting for me to turn on the camera for our Zoom sessions, I could hear their discussion and chatter. It was hilarious. They talked about the animals, the costumes, the movement, about what they thought I would

be wearing when I turned my camera on. They were someone else for that moment. Not a kid in lockdown.

So many parents told me that the kids loved art and
music during lockdown. One parent rang the school one
day in tears to send me a thankyou. She said her child had
been so unhappy and that this was the first time she had seen
him get moving and smiling that week.

My close friend Jacqui Dark wrote about the ability for
music to bond people. Jacqui faced many serious health issues
and obstacles in her life but her singing thrilled audiences
globally. She penned this after her first round of very brutal
chemotherapy treatment:

Singing and music absolutely got me through life. Whenever
I was devastated or in a terrible emotional place, I would
sing myself hoarse—it was a way to channel everything dire
into something that would somehow exorcise the darkness.
I've never been prone to depression or anxiety, so it was
summoned for those big moments in life when it all just got
too much. It works. It makes a difference. Music is a friend
when you feel you have none—a constant, all-encompassing
presence to turn to as an emotional outlet. I know it sounds
a bit extreme, but it's totally true. The professional musicians I work with invariably feel the same—music is more
than something they do, it's something they ARE, an innate
part of how they define their being. Nobody would do this
job unless they truly loved it.

Singing solo is visceral. When it works, you feel like
you're flying and connected to the whole universe. It's a

fusion of the physical, technical, emotional, tribal—the act of communicating with an audience and breathing as one is incomparable.

Singing as a part of a chorus/group is bonding beyond words. I will always remember singing as a part of my very first opera chorus, being for the first time surrounded by masses of brilliant voices and encased in a wall of sound that we were making on the spot. And that SOUND! You could feel the vibrations thrumming through you—yours and those of every other voice in the room. The fortes, the tiny piano phrases, everyone breathing together, everyone working as one to produce a joint creation. Indescribable (though I've tried!). An act of creation and connection as unique as it was ephemeral—existing only in that moment for all of us.

Music critic John Shand wrote about Jacqui's performance soon after a five-month Covid lockdown:

We were well into the concert when Jacqui Dark barely breathed, 'Ne me quitte pas.' Suddenly, after five months, you remembered why live music beats recorded music: you were no longer just another lost soul in a room, but part of a community of hundreds of throats gulping the same lump when Dark sang those words, and our presence made her feel and believe them all the more, in a potent spiral.[6]

The tendency for groups of people to synchronise rhythm even occurs when there is no deliberate intention or musical impetus. Bridget Farmer is a commercial artist. She wrote:

When I was at art college we did a two-week forging (as in metalwork, not art crime) project. We had to heat up our metal and beat it against anvils in order to shape it. It was a very loud project, with ten students all beating metal with hammers, but what happened was, after a few minutes of hammer hitting, all the students naturally synced up their beats. Our tutor, who had been teaching jewellery and silversmithing for twenty years, said it happened every time without fail.

David Kanosh talks about combining traditional knowledge and contemporary songs under the guidance of his grandparents in order to store traditional knowledge. This model can be transferred to storing any information.

As a child, my hand was the first loci used to teach me about myself.

Tip of my pinky finger is my Tlingit name (Yook'iskookeik).

Ring finger tip is my moiety (Raven). Middle finger tip is my clan (Deisheetaan).

Index finger tip is clan house (Shdeenhit/Steel House).

Thumb tip is clan of my father (Kaagwaantaan).

As I introduce myself to somebody of another clan I would say something like: My name is Yook'iskookeik. My moiety is Yéil (Raven). I am of the clan Deisheetaan. Our clan house is called Shdeenhit or Steel House. I am Kaagwaantaan-yadi (child of Kaagwaantaan).

Then my grandparents told me to look at our clan totem pole and pick out features which stuck out to me and use them as memory spaces for my identity which I then encode the information which I have previously encoded on my fingers.

My grandparents then told me to make a memory board and encode all that information relevant to my lineage and identity again.

The last thing my grandparents told me to do in this regard was to take the basics of my name, moiety, clan, clan house and my father's clan and put it in a song of my own composition or a parody. I sing those things to the song 'Don Quixote' from the musical play *Man of La Mancha*.

When I think back to my physics classes over the years, I see that there was always a talented musician or two among the students. There was always at least one student who was very active and there was always a natural leader. Of course, there was always a class clown, too. These are the students who will rally a class into song. I only wish I had taken advantage of their presence for us to sing and dance about electricity, mechanics, optics and all the beautiful formulas.

A jam session in the classroom

Of all the ideas I learnt from Margo Neale, it is her skeleton and flesh metaphor, mentioned in Chapter 3, that I seem to quote most often.

Every Aboriginal performance, she explained, must encode the information stored and convey it with accuracy. But it also must never be boring. And the solution is quite simple. All human bodies have exactly the same bones, arranged in exactly the same order, but no two people look the same. The bones of the knowledge are solid. The flesh each performer uses to encase the skeleton is his or her individual interpretation, and it will vary with every performance. Great performers are hugely admired partly because it is through them that knowledge is kept both accurate and memorable.

I think jazz offers a model that can take us way beyond great performances on stage. As mentioned in the previous chapter, music historian Ted Gioia emphasises that the traditional epics have emerged from our musical heritage. The only reason they have survived until the present day is because they have the support of rhythm and melody.[7] Ted explained to me that jazz musicians can play the same song every night, but each performance will be different. Jazz musicians have their favourite phrases, which they will use as building blocks within their solos. Great jazz players, he explained, will use their ingenuity to create something new in every performance. Within a jazz group, there is freedom to improvise, but the melody must still underwrite the performance. That improvisation allows a vast range of creative expression.

Jeremy Meaden is a professional musician. He plays trumpet in a traditional jazz group and in a rock band. He sees the jazz jam session as the model for group learning:

A jam session will generally proceed in a fairly well-established manner. First, the group of musicians will agree on a tune to play, or a tune will be called by the leader.

The group will begin to play through the tune together, with the 'rhythm section' (generally the drums, bass, and a chordal instrument like piano or guitar) playing the chord progression, while the front line instruments (which generally include vocals, saxophone, trumpet, etc.) will play the melody. Finally, once the playing of the melody, or 'Head', is completed, each member of the group will take turns playing an improvised solo over the chord progression, while being backed by the rhythm section.

This part of the performance gives the musicians their chance to 'flesh out' the tune on the 'bones' of the chord progression, and to put forward their own interpretations and ideas based around it. The rhythm section continues to ground the tune with percussive rhythms, bass notes and chords that are rooted in the original song.

These rules are of course not set in stone, and the circumventing, ignoring or reinventing of them in the performance is in itself an important part of the improvisation, but at the core of the jam session is the general knowledge that these rules exist, and that for a meeting of performers to achieve a meaningful, musical result then everyone must understand them well enough to stick to them, or to break them.

In the rock band, there is far less improvisation in a live performance setting. All of the songs are pre-planned and the solos are shorter and used as more of a short break between verses of lyrics. In the traditional jazz band, the music relies far more on the band's energy and reaction to each other in the moment. This also means that it can go terribly wrong, although that is what makes the music so exciting: risk and spontaneity.

During jazz solos to a jazz-literate audience, there is real audience involvement that can also inform the direction of the music, and the audience is an important part of the power dynamic within the performance space.

I see jazz as offering a valuable model for adding music to the curriculum. The melody—the core information from the topic—must be there and acts as a constraint, guided by the leader, the teacher. Each student can then improvise around that theme, creating a chant or, for the more ambitious, adding more complex music.

Every time you do something new, imaginative and interesting, such as creating a story or song for a piece of information, you also create new neurons from neural stem cells—that is, you trigger neurogenesis. Your brain loves novelty, and so much novelty is available if you pinch ideas from other students, teachers or anyone who does things even slightly differently. Copying when learning isn't cheating; it is expanding your experience.

To borrow further from a term commonly used among jazz players, Jeremy suggested that we all 'shed'. The term 'shedding' refers to going out to the woodshed to practise repeatedly and perfect a performance. In our private sheds, we can repeat chants, songs and dances of encoded knowledge without others judging our proficiency. A repeated refrain from the Advisers is that learning is all about the process, not the final product. It is in your shed that you can develop the improvisations which add flair to the basic knowledge held in the melody.

In my shed—usually the shower—I sing loudly and out of tune, and add lots of trills and extra bits. I love it.

Art adds a different way of engaging

For Alice Steel, the intersection between art and science is part of her heritage, with family expertise in both domains.

She is now a qualified biological scientist and a practising artist. She explains:

> I often had to choose through high school: do I do art or do I do chemistry, drama or maths? I had to choose between them because of conflicts in the timetable. Either you're in the science and technology stream or you're in the art stream. We need to find a fusion of the two.
>
> When I ran kids' art classes I'd find ways to include science, and then when I ran science classes I'd try to infuse some art, because that is my nature.
>
> I did one class about luminescence, where we looked at lots of interesting things that glow, including live scorpions under ultraviolet light. At the end we did art with fluorescent paint and shone it under UV lights. Their artwork glowed—a big wow factor. I knew that they had really engaged with the subject matter because their artworks were annotated with things like 'we love science' and 'this is the best class ever'.
>
> That's what science needs. Science communication needs that infusion of art and having fun, which hooks them in and tells them that science isn't something they need to be scared about or that it's some hard technical thing that only really smart people can do. Everybody can do this and explore and play around with experiments and if you add that fusion of performance and art, it captures everybody.

Tens of thousands of years ago, the earliest artists left panels of abstract symbols interspersed with recognisable images of people and animals and plants, of the skies and of ceremonies. Their descendants over the millennia continued the pattern.

They almost certainly used these panels as memory aids for their songs and stories.

Nearly a thousand years ago, someone envisaged the Bayeux Tapestry, which isn't a tapestry but a 70-metre-long embroidered cloth. It tells the story of the events leading up to the Battle of Hastings in 1066 and the Norman conquest of England. Made in England, it was probably embroidered soon after the story it tells.

A community-generated knowledge panel would work a treat in any communal learning space. The simplest way I can imagine doing this is with any form of cheap, strong paper sold in large rolls. One Advisers' meeting became quite animated as we imagined a classroom creating their 'rock art panel'.

At the beginning of a new topic, a clean, inviting, empty length of paper would be placed along a wall. As each critical point is made within the curriculum, students can annotate the panel with images and symbols representing what they are learning.

Students with strong artistic skills might wish to display their flair, much like those sweating away at the incredible horses and the bison in Ice Age caves. Others might add stick figures acting out events, or include abstract symbols the whole group can identify. Those inclined to rhythm and rhyme might create an ongoing ballad, adding verses or lines as the discussion develops. Class members can tell the stories, relate the knowledge and eventually see an overview of the whole topic, using that panel as their prompt.

At the end of the topic, students could photograph the massive mnemonic narrative. Replicating the panel at a personal level would serve as a very robust revision.

The reason we decided to use a length of paper rather than some more permanent fitting within the classroom, such as a whiteboard, is that the teacher can roll it up at the end of every class to keep it safe. We are all realists.

Sometimes, a surface is available to create a permanent artwork. In all cultures, learning about ethical expectations is critical. Paul again:

In 2022, I secured funding to paint the concrete play pipe in the playground. It was to teach the four school values: respect, honesty, inclusiveness and integrity. On the outside of the pipe, the kids had to represent what it looks like when someone was being respectful or someone was being inclusive. On the inside of the pipe was the internal: how does it feel when you're being inclusive or respectful? They came up with gorgeous symbolic art forms for it. For example, for honesty, on the outside they painted a key with a rainbow behind it. According to the girls who made this design, the key was the key to everything. On the inside was the sun shining through the clouds, and the clouds clearing.

The children took six months or so to come up with the various designs, working in pairs, doing a lot of drawing. They started with things like two people standing next to each other being respectful because in their mind they were almost seeing movies of two people dialoguing with one another. It took a lot of process of boiling it down and boiling it down and making it more and more, not abstract but simplified and concise, for them to put that into a still, simple, static image. That boiling down process reduced and reduced and reduced to get the gold at the bottom. With some of the girls, it is still two figures there for honesty, but it's a girl on one side with the devil over her telling her to be dishonest and the other one has an angel over her telling her to be honest. This symbolism evolved from their interactions with one another.

The teachers refer to these images when explaining to kids what it is to be respectful. Each interprets that differently. The picture doesn't change but the meaning and the way that meaning is constructed for the child might be different, but we can all relate to the symbol. Eventually the paintings will fade because they're out in the hot sun. Then, just like Indigenous cultures, we can repaint them as we teach the concepts.

That artwork is within the students' experience every day, reminding them of the lesson learnt. To implement these ideas in schools will take an integration of staff skills and, for some teachers, a significant change in mindset. We'll need innovators like Paul.

It is not a great leap from a massive panel on a rock face to see the same cognitive skills employed when Australian Aboriginal cultures draw their stories on the ground in 'sand talk', produced as the stories are told. Most classrooms and homes do not have a suitable patch of sand, although I think this would work a treat if an equivalent could be found.

However, there are other permanent models to imitate. The Anishinaabe tribes of North America inscribe scrolls made of the bark of birch trees, with abstract images acting as memory prompts for their stories. In the film *Wiigwaasabakoon/Birch Bark Scrolls* (2011), scroll keeper Miskomin teaches Indigenous

American students how to create a birch bark scroll telling their contemporary stories and using their own symbols. Miskomin expressed the belief that 'this kind of form of communication is going to be important again'.[8] I think that time has now come.

I created a simple scroll on a rolled-up strip of paper in very little time, telling stories of the planets of the solar system. This was a sample for a lesson for Grade 5 students, around ten years old. After giving each planet a human form, they are shown arriving at a party given by the Sun. They turn up in planetary order, each adorned with symbols recording details of their moons, orbits and environments. Saturn was covered in rings and Jupiter appeared to be in a particularly stormy mood. Turning the planets into humans makes them so much more memorable. Stories depend on a compelling cast. Even the most abstract concepts become memorable if you turn them into characters.

I have watched teachers in workshops create scrolls for their subject areas, everything from psychology to Shakespeare. The cost involved is minimal.

The narrative scrolls of China and Japan offer a more extensive model. As I talked about in Chapter 10, these glorious objects tell stories in a format that is compact and easy to store. I have a replica of the Chinese handscroll *Spring Morning in the Han Palace*, a precious treasure. It is over five metres long, designed to be gradually unwound to reveal only one portion at a time as the story is told.

Qiu Ying (*c.* 1494–1552) took up his brush to depict the story of one of his fellow artists working in a Han Dynasty palace, around 1500 years before. The long scroll shows various activities in the palace, including the famous story of Mao Yanshou (毛延壽) painting the portrait of Wang Zhaojun (王昭君), as shown in the portion in Plate 11.1. With numerous concubines, Emperor Yuandi (r. 48–33 BCE) instructed Mao Yanshou to paint their portraits for him to choose his company. The concubines bribed Mao to exaggerate their beauty. Wang refused to do so. Consequently, the artist's portrayal of Wang was less beautiful and she was never chosen. In settling a dispute with a barbarian chieftain, the emperor offered the woman he considered the least attractive of his concubines to the chieftain as his wife. When he saw Wang, Emperor Yuandi realised that she was the most beautiful of all. The emperor had the artist executed.

Narrative scrolls are a tradition in Korea and India as well, often portraying stories from the great Hindu epics, including the *Ramayana*. But it is from Japan that we can continue the storytelling sequence. As we saw in Chapter 10, the Chinese narrative scrolls are part of an ever-changing art tradition that dates back to the rock art panels. These were adapted by the Japanese (see Plate 11.2) and eventually ended up as a comic format known as manga.

Planning and then laying out our information while imaginatively using the space on a page or scroll is a creative process that aids recall immensely. We are naturally disposed to remember using spatial arrangements. Every format in the continuous art story from the earliest marks offers a model for the classroom or individual learning.

Commercial artist Bridget Farmer works in a range of media and for a range of purposes. Plate 11.3 shows one of Bridget's book illustrations on the left and an 'art' version on the right. The illustration puts a spotted pardalote in context and uses accurate colours. The artwork, of a superb fairywren, tells you far more about Bridget's response to the tiny bird's personality.

She described her thinking to me as follows:

The first image is an illustration from my children's book *The Bush Birds*. I am trying to depict the bird, still as an artist, but I also need it to be easy to recognise and maybe use as a reference if using the book to identify birds. The colour has been added using watercolour paints, something I don't do with my fine art work, and I feel the illustration is a bit tighter than my other work.

The second image is an example of my printmaking work, my art for art's sake. I was striving to be looser with this piece. I fight with myself to try to keep lines loose when working on this kind of art. I do find I constantly tighten up, which works okay for the illustrations, but I want to break out of that for this sort of work. However the medium of drypoint does make it difficult to be loose due to the pressure I have to exert on the etching needle to create the scratched lines in the metal. I'm trying to re-create effortless mark-making with a medium that requires much effort to make the marks!

It might not actually be that obvious, the difference between my illustrations and my non-illustrative art, to the outsider; maybe some pieces are more so than others. I think it's sometimes just my mindset while making them rather than the end result and how I'm thinking about who the audience will be.

Jane Rusden is the dyslexic artist we met in Chapter 6. She and Bridget talked a great deal at our meetings about the potential for art in learning, something I am only glimpsing now as I try to follow their lead. Jane wrote:

If you don't have a deep understanding, if you don't fully understand what you're painting, you can't represent your subject.

Being an artist, you need to be a very good observer. You need to be able to see, you really need to SEE. You need to be able to observe the world around you and understand the ebbs and flows and understand what underlies everything in order to depict it successfully.

The way that I bring narrative into art is through my subjects, so my subjects need to deliver dynamism. I don't just want to paint a bird on a stick like a lollipop. That doesn't have a story for me and it's not terribly interesting. To create narrative I need the bird to be in a particular position that says something about its physiology, its behaviour, about where it sits in the ecology of the world. Composition is really important. By leading the eyes through an image, you can tell a story as the viewer's eye goes through the picture.

Jane's bird paintings are constrained by the reality of the bird and the story she wants to tell. Her style allows the story to be told with her individual flair. The painting shown in Plate 11.4 hangs in our living room. Although we have other paintings of a similar size, the falcon dominates the room, just as the real bird does its environment.

Representing ideas in images, illustrated within the constraints of the knowledge content, forces a focus on the topic at hand. You have to notice the details, and that imperative is so wonderful for learning.

Memory boards work a treat

The memory boards discussed in Chapter 4 transfer wonderfully well to learning for any age. I get more emails from people who have created memory boards than I do about any other topic. Their excitement is always palpable. Their shock at experiencing the effectiveness of memory boards is universal. The impact is long term.

I use the same method in classes with students from the age of about eight to adults. Children are usually working to a topic defined by the school curriculum. Adults usually choose their own topic. I explain the background, at the appropriate level for the class. Laid out on a long table are bags of beads collected from discarded jewellery donated to a charity shop. The cost is minimal. Each student takes a board of soft craft wood and heads for the tables. At that stage, every class goes silent, even though that is never requested. Students, young or old, search the table intensely for the exact bead to represent each concept. In one class, the silence was broken by a shout of joy when a woman found a red heart-shaped bead, perfect for her board about the circulatory system. The process of choosing just the right bead consolidates the concepts within the topic.

The process of arranging the beads on the memory board consolidates the relationships between the concepts. And then the class tends to get noisy as neighbours share their finds.

Even when they are all doing the same topic, the boards end up looking very different, just as is reported for the African Luba when using a *lukasa*.

This individual interpretation of the same topic has been particularly obvious in a regular workshop at Melbourne University. Duane Hamacher, Associate Professor of Cultural Astronomy in the School of Physics, uses memory boards when teaching constellations. Students choose their beads, sometimes grouping more than one to represent a constellation. The first year, some students stacked them. I had included a few objects from the discarded jewellery such as feathers and bits of chain. These were so popular that the next year I had a lot more variety on offer.

I was delighted to receive this message from Duane: 'I'm grading the memory videos of students using the *lukasa* to memorise the 88 constellations. You should see them, they are amazing. They are all really engaged and excited. They are supposed to be 10–15 min videos. One video was 32 minutes.' A student from the workshop contacted me later. He was making memory boards for his classes in the medical school, and was delighted with the advantage they gave him.

Duane turned up to one workshop with Ghillar Michael Anderson, an Aboriginal Elder, Senior Law Man and leader of the Euahlayi people. Although the workshop is based on the *lukasa* of the Luba people, Uncle Ghillar confirmed that the cognitive process was just like that for using a restricted Aboriginal object, the *tjuringa*.

Alice Steel has been teaching memory workshops for children as young as three through to adults. She told me:

You need the creative tools to help you embed science so much better. In a science class with young teens, we made little *lukasas*. They were incredibly effective. Within just an hour and a half, the students were all reciting the scientific names of all the acacia plants in our area. I did the Victorian parrots with a different class. They sculpted clay eggs with symbols for all the parrots on them. They could recall all the parrot names.

It worked for me too. To this day I know all of those acacias and parrots.

One of my favourite memory devices is for the native orchids of our region. It's a flat clay disc that I've carved symbols into. I love the feel of it. I love the beauty of it and I love the way it holds some of my brain. I don't 'use' the physical object anymore, because the knowledge I embed by making it is in me now. I'm not able to *not* use them anymore. The picture of that device comes into my head when I'm trying to identify orchids and it's the anchor for that knowledge.

Learning is so much better with theatre

All Indigenous cultures involve performance-based knowledge systems. As a teacher, the more theatrical I became, the better my classes went. I now regret that I did not take that far enough. It is one of my concerns about the trend to open classrooms. I would not have been anywhere near as exuberant with other teachers in the room or with less control over the space.

Alice Steel is now a science communicator at the Discovery Science & Technology Centre in Bendigo, Victoria. She takes full advantage of her artist thinking when presenting science, as she explained to me:

As a scientist, as a science communicator, I could just spew all the facts of things. When running science shows, we could just stand there in a lab coat and present the information. It would be incredibly boring.

In a twenty-minute show you want whiz, bang, pop, hydrogen explosions, liquid nitrogen, exciting physics and chemistry . . . but what I really wanted was to design a biology show. Biology is hard to demonstrate because it's slow.

It takes a long time to portray life cycles of animals, or plants growing from a seed into a full-grown tree. But I knew I could do it because I'm also an artist.

The theme of the show was 'Matter of Life' and it took the story of matter from the Big Bang through the formation of the planets and the emergence of life. We actually pretended to make primordial soup on stage, adding in that performing element. I said to the staff who were running the show: 'Act like you're making a really delicious soup. Add in a bit of spice, a sprinkle of nitrogen, a dash of calcium, some oxygen, etc.'

Then life emerged. We used lots of different animal puppets and the kids had fun working out which ones were carnivores and which ones were herbivores and learning about the transference of matter from one life form to another through eating.

I sewed a life-sized prop of human intestines, with lovely elastic in it. It would start all bunched up inside a skeleton, called Gary. I'd ask, 'Gary, you don't mind if I borrow this do you, seeing as you're not using it anymore?' Two volunteers, from the audience, stretched it out to 7.5 metres. Yes! Actually, on average everyone's intestine is that long. And people go, 'Oh my god, that's full on.' Obviously I couldn't bring in a real intestine without grossing people out, but the act of them stretching the prop, experiencing it, seeing it, the art element, really uplifted the learning experience for those kids.

And then I couldn't leave out what happens after digestion. Of course, we end up with poo, getting the expected giggles. This is good. This is people emotionally engaging.

We had a coprolite, a fossilised dinosaur poo. Everyone got very fascinated with that, but most poo doesn't end up as a fossil. It gets churned up and eaten by creatures, worms, dung beetles, and turned into soil. It decomposes.

My problem was how to end the show. I couldn't just end with dirt and death. I wanted to end with renewal, rebirth, and somehow a really big pop-bang finale. Seeds are a good way to talk about reproduction, and growth from that soil we just talked about. So, how to show a plant growing and fruit appearing at the top?

I got an old bicycle pump and attached a long tube to it. I wound green tape along the tube and made these beautiful leaves, turning the whole length into a vine. We could slowly hoist it up with a pulley so it would slowly grow out of a pot. At the very top, I attached a balloon, deflated. I managed to track down those seeds that float and spin as they fall. Several of these were packed secretly inside the balloon. A kid from the audience would come up and use the bicycle pump to slowly grow the fruit at the top, bigger and bigger and bigger and so big that the audience was impatiently waiting for the balloon to burst. Perfect for building up the suspense and tension at the end of the show. And then, pop! And sprinkle, sprinkle, sprinkle, all the seeds flutter down. This arty prop allowed people to participate in the growth of a plant, and allowed us to come full circle from the Big Bang at the start of the show to a big bang at the end. The seed dispersal showed that matter and life would continue on.

Teachers don't always have an Alice available to make props, but introducing a theatrical element to teaching, as many good teachers do, makes a huge difference to learning outcomes.

Let's get practical about implementation

Dream as I might that we can all suddenly take advantage of the untapped potential that the knowledge gene offers, the reality is a bit more complex. Paul explained what happens when he teaches the multiplication tables through drawing:

> When we did multiplication tables this year, the kids did all the stories in literacy class in Term 1 and completed the drawings in art all through Term 2. But their efforts were not taken up in the maths class. There is still silo thinking in there. The classroom teacher was not taking advantage of what the children knew.[9]

What we need is the maths teacher using art, not the art teacher teaching mathematics. Paul saw another problem with this concept:

Teachers are traumatised by the whole 'I can't even draw a stick figure' type of thinking. We don't teach kids to read and write just so they can grow up to be novelists. We teach them to read and write so that they can learn and fully participate in our society, which is very literate based. Art is the same. We learn lessons through art that can't really be learnt otherwise: observation, dexterity, imagination, creative thinking, critical thinking, the capacity to be able to disagree because happy is orange for one kid and blue for another—neither is right or wrong. We just have different tastes and opinions. It's a huge lesson. Most adults need that lesson, let alone children.

Most people have to overcome the idea that drawing stick figures is somehow inadequate. We can all think representationally. Whether or not we can actually accurately portray to the level of Michelangelo or Rubens on a page is immaterial. Research has shown that if you draw your shopping list, you are more likely to remember it than if you write it. I tested it out. Half of the class wrote out twenty words, *cat, hat, boat* . . . for a minute while the other half drew the cat in the hat on the boat in a minute. I tested them all and those who wrote each word down as many times they could in the minute would score maybe 14 out of 20. The ones who drew it scored up to 18 or 19 out of 20.

Paul also talked about the difference, when teaching critical thinking, between having a child explain an artwork and reading or writing text about it:

What do you see in this painting? What do you think is happening in this painting? What do you think the inter-action is in this painting? Okay, why? What are you basing it on? Many times, the child will notice something in the painting that I haven't noticed. Ask a young child to sit down and write something, and it is sometimes torturous to see them with their slow handwriting. They can't com-municate the thought at the pace at which it is happening. When it's a verbal dialogue, as it is in the art room, it's at the pace of the thinking, so kids can't lag behind or have to drag their feet waiting to get it out. It can happen at the speed of thought.

'Modality shifts' occur when you take information pre-sented in writing, say, and shift it to another mode, such as a chant, illustration, story or song, or when you work out how to encode it in a memory palace. Creating a mode shift forces concentration, the enhanced attention that the knowledge gene so positively assists. Once the basic theme of any topic is grounded in memory, this builds a firm foundation for the higher levels of learning and thinking to flourish.

I am not suggesting every classroom be transformed into a creativity powerhouse for every topic. It is asking way too much of teachers to have skills and resources in every one of the technologies I've talked about. It would be naive to think that is even remotely feasible.

But I do think there is a way this can be done, if the school, university or other education institution has the resources to unleash the full potential offered by our knowledge gene.

Music and art programs in schools and beyond have an important role in supporting talented individuals headed for the stage and galleries, and those courses need to stay in place. But then there's all the rest of learning. We can also bring music, art and memory palaces into the heart of all education without changing the basis of the curriculum.

I was part of the introduction of computers into schools many decades ago. Information Technology was a new subject. Gradually, over the following decades, computer skills were integrated right across the curriculum. Some technology-savvy individuals took specialist classes. The rest just took advantage of what the technology had to offer their field. I see the same model for music, art and memory palaces.

Just as we have IT technical support for teachers and students in schools, so there should be music, art and memory palace technical support. The Knowledge Hub Advisers should be available to advise teachers, help in classes and provide access to portable art and music equipment, which can be taken to classes or to the locations in the memory palace.

There may be support staff already who have skills in these domains, or there may be room for a new staff member. Every institution will be different. Schools I have worked in have always had eager parent and grandparent volunteers. Many will have invaluable skills. There are also many people in the community with art and music skills, a potential volunteer base that goes way beyond the parent body.

Damian has worked in many fields. (All authors should choose partners as useful as mine.) His career in education included teaching commerce, running libraries and heading

IT support crews at both school and university level. Here's his view on this topic:

> Based on long experience with school libraries, I know that operating an open, accepting library provides a comfortable place for divergent kids to spend time where pressures to conform are reduced. With a free-wheeling approach that accepted differences it often gave recognition to kids who were sometimes considered odd. Such a place can easily be enhanced with art, music and technical support people. It is not a place of conformity but rather an open and accepting environment. Unlike many areas of a school it is also a place where different age groups can interact.
>
> These days, municipal libraries are often at the forefront of community education because of the broad range of activities they encourage.

Many community libraries already offer technical support for computing to a diverse clientele. They have access to a broad range of people, who can be mobilised into a volunteer base for music and art. A permanent memory palace around the library would act as a model for how to use familiar spaces for a set of locations.

We have a population in which there is a screen in every hand. I want a musical instrument and a paintbrush to be just as enticing. I want everyone walking their environment and setting up memory palaces without earphones blocking the outside world. I want people using their own imaginations, creating their own stories and increasing their own knowledge. I have no doubt that this can be done—and that it should be.

Potential for learning lasts a lifetime

Taking advantage of our genetically driven skills is not something that is available only to the young, however you define that word. Our genetics rule us for every moment we breathe.

Learning can last a lifetime. When finished with school and tertiary training, our brains are still primed by the knowledge gene to learn. We can read and write and listen and watch to learn. But by also engaging with music, art, memory palaces, narrative and performance, we can do so much better. It is so easy to take advantage of these enhanced learning arts in workplaces and community centres, in social groups and aged care homes. But most of all, our own homes are the perfect setting for engaging with all that the knowledge gene offers throughout our lives.

At the time of writing, Peggy Shaw is 101 and a half. (She insists that, at her age, the half is very important and I must include it.) Peggy has been a commercial artist for most of her life. She still paints every day, and sells her work both as originals and through a commercial card company. She told me that she can't die yet—she has too many more paintings planned.

In this book, I have waxed lyrical about the role art can play in a knowledge system, as a superb memory tool. But it is important to acknowledge the validity of art for aesthetic purposes only. Peggy considers the images she creates are purely for the beauty of the form and representing the flowers she loves so much. That does not mean she has ceased to use the visuospatial abilities that her knowledge gene provides. Peggy takes a three-dimensional vase of flowers and produces a two-dimensional image of it. In her mind she is visualising and transposing.

As someone now in my seventies, I am delighted that current research on brain plasticity indicates that it doesn't have to decline with age. We can keep on generating new neural pathways throughout a life as long as Peggy's. The joy of art keeps her wanting to 'work' every day. And throughout that very long life, we can take full advantage of what the knowledge gene offers.

So much of the political and economic talk seems to be all about endless growth, while advertisers bombard us with the need to buy and buy and buy even more. Since connecting with artists and musicians and storytellers, I have become aware of a totally different view of success. As Alice Steel told me:

> It's not about pay when you do art. Mostly it's about a means of expressing yourself, it's therapy, it's communication, it's a way of having a voice in a different flavour. You can't just say something with words all the time, but if you can paint it, draw it, perform it, sing it, dance it, sculpt it, photograph it from different angles, you get way more perspectives on an issue and engage more people.
>
> Society needs to value that creativity more, because it enriches every element of our lives. It sparks the wonder, the 'what if' that drives us to invent and discover, and it's foundational to the way we play, learn and adapt. Imagine architecture, the clothes you wear, the technology you use, entertainment of any kind. Without that element of art it would be very dull. The Arts make our lives interesting, make life exciting, make us want to interact and start conversations. Make us want to be involved.

We are all biologically programmed to use the skills granted by our genome. Modern society the world over has disparaged the use of the combination of knowledge skills that evolution ensured was embedded in every one of us. The knowledge gene and the language gene, influenced by other genes, have worked together to give us a powerful information system for all of human existence. We use the language gene to its full potential. If we can recoup and optimise the skills our knowledge gene offers us, while we keep talking, reading and writing, surely this is the most optimised system of all.

We are genetically encoded to use a combination of music, art, story and a connection to place in learning and knowing, but we have gradually neglected this as literacy took hold. We can have it all.

WHERE TO FROM HERE?

Nearly two decades ago, my life became so much richer when I started exploring the memory systems used by Indigenous cultures. My lifelong problems with memory slowly evaporated. I felt that I had found a sort of paradise where I could finally learn anything I wanted.

I reluctantly relegated all the projects incorporating my new memory techniques to the backburner when I received Andrea's extraordinary email three years ago. All the doubts and scepticism I had on the first reading have now evaporated. The research conducted with Andrea and Vic gave me an awareness of the skills I have failed to value throughout my life—skills innate in everyone reading this book.

As I return to my shelved projects, I know that the three-year hiatus has granted me a much broader, deeper and more beautiful capability with which to attack my favourite activity: learning. I return with a nuance and confidence that didn't exist before I knew of my double dose of NF1, my genetically encoded benefit. I feel as though I am at the start of a new journey, conscious of my knowledge gene as well as gaining an appreciation for those who think and behave differently.

I am also much more in touch with my immediate environment. I notice the plants, trees and grasses. I notice the Lilliputian animals acting out their lives on tiny stages. I can't help but recognise the way everything is interdependent. Something has changed in my awareness of the natural world. This delight is available even to those who live in the hearts of cities, where that plant might be a tiny weed and that animal a buzzing fly. Everywhere, there are natural sounds and rhythms that reveal themselves when you stop and listen.

We need connection to place. We need to feel that connection to every detail of the mesh on which our very existence depends. Our planet is inconceivable as a whole, but it is just billions upon billions of tiny memory palaces. And every one of them needs our protection. We cannot learn that from dull, written reports and stagnant presentations stuffed with data that have no story, no art and certainly no music.

The music that was once in the background of my life has now moved into the foreground. Before this research, I hadn't understood the social and emotional impact of music. I have a tendency to overreact. My family will laugh as they read that understatement. Now, when I need to calm down, I play music. When I am depressed or lacking motivation, I play different music. There is now so much music in my life.

I no longer restrict myself to writing as my only communication tool. I am drawing and painting and developing long scrolls that tell the stories of newly gained knowledge.

Through the insights granted by the research instigated by Andrea and Vic, I have reflected on decades of teaching. We need to teach students as fully human students—diverse, emotional, biologically driven students. We need to offer our students the chance to be immersed in memory palaces brought to life with music and art and incredible stories.

We need to take advantage of all the potential we are neglecting by undervaluing our genetically driven cognitive vastness. The world desperately needs divergent thinking.

We need to reinvigorate our engagement with art, music and story, placing them at the heart of our lifetime of learning.

Our human NF1 allele embeds in us the potential that evolution favoured so strongly, and we need to take full advantage of this untapped potential within our communities. Evolution also ensured that some of us were born different—diverse— offering a breadth of ways of knowing that works for the whole population. We are doing a massive disservice to our future if we don't embrace the diversity that served humans so well over tens of thousands of years.

I want to continue to work with those whose words you have read throughout this book. So much of their wisdom could not be included, so the research will not stop here. There is much more to do by capitalising on the neglected skills the knowledge gene reveals, not only in education but also in the way we think about our local communities, wider societies and the world's population as one humanity.

I acknowledge that this is all far easier to write than to implement. It requires a change in mindset from a world that values wealth and power for the individual to a world that values the strengths and needs of the whole population. It requires us to stop seeing endless growth and profits as worthy goals if they come at the expense of the planet and of so many people on it. But we can change. We can. We need to do an honest cost–benefit analysis, just as evolution did when granting all of us two copies of the knowledge gene.

I still can't sing in tune, but I can sing because we are all musicians. I can't create the artistic masterpieces I admire in galleries, but I can produce art because we are all artists. We can all tell stories, and relish those told by the best storytellers. And we can immerse ourselves in the physical environment in which we live and on which all life depends.

Our uniquely human skills, which evolution ensured were innate and universal, are life sustaining. We can all take more advantage of what it is to be quintessentially human. We can embrace the ancient skills our society has lost or sidelined over the last few thousand years. In doing so, we can become more fully human. Together, we can change the world.

APPENDIX

The knowledge gene skill set

The following 89 skills were identified in the knowledge systems of oral cultures, divided into seven categories: Music, Performance, Spatial, Art, Language, Cognitive and Social. Those that were selected are highlighted in grey.

Key:

†	challenges for the NF1 cohort
★	skills observed in chimpanzees
NSD	no significant difference
PS	possible strength
UN	unknown/no relevant testing

Music skills

1. Perform music in songs or chants to store knowledge[t]
2. Use rhythm as a memory aid (mnemonic)[t]
3. Add complexity to music[t]
4. Create embodied musical forms[t]
5. Encode knowledge in dance[t]
6. Use music to generate emotional response[UN]
7. Compose music[t]
8. Create musical instruments[UN]

None of these skills have been seen in chimpanzees. Skills 1, 2, 3, 4, 5 and 7 were assessed as being a challenge for those with the NF1 disorder, especially given the high rate of amusia. Embodied musical forms include all forms of rhythmic movement to music, which includes the critical role of dance. Skills 6 and 8 could not be rated as there has been no relevant testing done.

Performance skills

9. Enact knowledge in performance[UN]
10. Take on different persona in performance[NSD]
11. Mimic animals in performance[NSD]
12. Represent plants in performance[NSD]
13. Predict and enact possible future events[NSD]
14. Maintain core pedagogy accurately, embellish for entertainment[t]
15. Sequence a performance of multiple elements[t]

Some semblance of skill 15 has been observed in chimpanzees, as male displays can incorporate performances involving a sequence of elements. But they have never been observed creating the complex performances with multiple participants that are so common in human ceremonies, so this skill has been retained.

There is no research to indicate whether or not skill 9 would pose a challenge for those with the NF1 disorder, so it was eliminated. No significant difference has been observed for skills 10, 11, 12 and 13. However, skills 14 and 15 were considered to be a challenge for the NF1 cohort due to issues with executive function.

Spatial skills
16. Conceptualise vast areas of land[†]
17. Use structured knowledge of land as mnemonic[NSD]
18. Visualise distant spaces[†]
19. Navigate from information in songs[†]
20. Conceptualise a constantly moving skyscape[UN]
21. Link the skyscape to the landscape[UN]
22. Use cardinal directions rather than right and left[†]

Chimpanzees display some spatial awareness in knowing the location of resources within their territory, as required for skill 16. But there is nothing that would indicate that chimpanzees encode those vast spaces with stories, conceptualising them as knowledge spaces. They do not mark specific sites or in any other way indicate the significance of their landscape. Hence skill 16 was retained. None of the other spatial skills have been recorded for chimpanzees.

For those with the NF1 disorder, skill 17 was assessed as showing no significant difference. There is no research relating to skills 20 and 21, although it can be predicted that these would provide challenges due to visuospatial problems.

It is the difficulties those with the NF1 disorder are likely to experience with skills 16, 18, 19 and 22 that are highly significant for this research.

Art skills

23. Create art primarily as mnemonic, not aesthetic[NSD]
24. Visualise objects in 3D[†]
25. Use art as representation of place[NSD]
26. Create abstract art as mnemonic[†]
27. Use abstract elements to enable adaptability[†]
28. Use ephemeral art as learning process[NSD]
29. Produce standardised designs[†]
30. Adhere to standardised designs[PS]
31. Create representational art[†]
32. Organise art in ceremony[†]
33. Participate with art in ceremony[PS]
34. Create art communally[PS]
35. Create pictographic art[NSD]
36. Adapt established artwork for new knowledge[†]
37. Utilise manual dexterity[†]
38. Create memory devices on wood, stone, cords or fabric[NSD]
39. Create art on utilitarian objects[NSD]
40. Use art as a passport, a message to other tribes[NSD]
41. Use poles, posts and tree trunks for mnemonic art[NSD]
42. Combine art with performance[UN]
43. Use natural colour sources[NSD]
44. Visualise a new form from common materials[★NSD]
45. Create tattoos to standardised forms[NSD]

Chimpanzees have not been observed demonstrating any of the art skills other than skills 37 and 44. In grooming, nut-cracking and fishing, they demonstrate fine motor skills, but none of the manual dexterity required to create and manipulate complex tools and artworks. Although chimpanzees do modify objects to use as tools, they do not create objects that require carving wood, chipping stone or moulding clay as humans do. The complex manual dexterity of humans justifies the retention of skill 37, while the three-dimensional visualisation is already accounted for in the retained skill list because of skill 24.

Those with the NF1 disorder find challenges with mastering tasks requiring visuospatial abilities or fine motor skills. Skills 24, 26, 27, 29, 31, 32, 36 and 37 were therefore assessed as providing challenges for this cohort. Skills 23, 25, 28, 35, 38, 39, 40, 41, 43, 44 and 45 were assessed as indicating no difference from those with two fully functioning NF1 genes. Skills 30, 33 and 34 were assessed as possible strengths for those with the NF1 disorder.

Skill 42 couldn't be assessed as there is no relevant research to draw on.

Language skills
46. Prosody*
47. Work with language in complex forms[†]

Chimpanzees demonstrate prosody in their different vocalisations. Due to speech challenges for those with the NF1 disorder, skill 47 was considered a challenge.

Cognitive skills

48. General comprehension[*NSD]
49. Repeat information extensively and systematically[PS]
50. Retain an integrated knowledge system[†]
51. Use narrative and mythology as mnemonic[†]
52. Use metaphor to reduce mnemonic effort[†]
53. Optimise focus, concentration and perseverance[*†]
54. Layer knowledge in levels of complexity[†]
55. Maintain classification of aspects of the environment[PS]
56. Optimise memory skills[†]
57. Act upon new observations encoded as rules[PS]
58. Integrate environmental knowledge[†]
59. Distinguish between reality and fantasy[†]
60. Conceive inanimate objects as living[NSD]
61. Curate and utilise apparently non–utilitarian objects[NSD]
62. Conceptualise time and maintain a calendar[PS]
63. Gain knowledge from experimentation[*NSD]
64. Use imagination to create characters and scenarios[†]
65. Exploit anthropomorphic thinking[NSD]
66. Make patterns with common objects for memory aids[†]
67. Add new classifications of environmental aspects to existing system[†]
68. Encode knowledge efficiently[†]
69. Decode knowledge from encoded formats[†]
70. Maintain complex genealogies[†]
71. Value knowledge for knowledge's sake[NSD]
72. Think scientifically[*NSD]
73. Record observations and act upon them[†]
74. Exploit magical thinking[NSD]
75. Create structured, coherent knowledge formats[†]

It is important to note that those with the NF1 disorder were assessed as having no significant difference from those with two fully functioning NF1 genes in general comprehension (skill 48). The challenges they experience are not related to intelligence.

Chimpanzees show some degree of general comprehension. They also demonstrate skill 53 in their focus, concentration and perseverance in tool use. They gain knowledge from experimentation (skill 63) when devising tools and demonstrate some ability to think scientifically (skill 72).

There is no significant difference in skills 48, 60, 61, 63, 65, 71, 72, and 74 between those with the NF1 disorder and those without. Skills 50, 51, 52, 53, 54, 56, 58, 59, 64, 66, 67, 68, 69, 70, 73 and 75 were all assessed as presenting challenges for those with the NF1 disorder. Skills 49, 55, 57 and 62 were assessed as possible strengths due to a preference for rules, schedules and repetition.

Social skills

76. Ensure adherence to formal teaching processes[PS]
77. Ensure public/restricted knowledge dichotomy maintained[PS]
78. Ensure adherence to public and restricted ceremonial places[PS]
79. Maintain structured knowledge formats[PS]
80. Adapt to community needs★[†]
81. Enhance cooperative skills★[†]
82. Organise ceremonies and cycles[†]
83. Participate in ceremonies and cycles[PS]
84. Record and ensure maintenance of processes[PS]

85. Use charisma and physical presence to maintain audience interest★†
86. Maintain law through socially accepted punishment★†
87. Maintain a cohesive social structure with material egalitarianism★PS
88. Use emotional intelligence†
89. Maintain relationships with humans and with others★†

Chimpanzees are social animals. Although not to a human level, they were assessed as demonstrating skills 80, 81, 85, 86, 87 and 89.

Those with the NF1 disorder were considered to find skills 80, 81, 82, 85, 86, 88 and 89 as possible challenges. They were also assessed as demonstrating possible strengths in skills 76, 77, 78, 79, 83, 84 and 87.

The social skills could not be considered indicative of the impact of the NF1 disorder on the human NF1 allele and the category was not included in assessing NF1 as our knowledge gene.

ACKNOWLEDGEMENTS

First and foremost, I must thank Dr Andrea Alveshere for both her discovery of the unique human allele of NF1 and for making the link between this gene and my work on Indigenous knowledge systems. Without Andrea's insight, I would not have had such an exciting new lens with which to revisit and expand upon my research. I must also thank Dr Vincent Riccardi for his involvement in this research and his unique and expansive knowledge of neurofibromatosis.

Elizabeth Weiss, my publisher at Allen & Unwin, has been closely involved with this project from the start. Her advice has been invaluable and her support constantly reassuring. I am indebted to Managing Editor Angela Handley for her expertise overseeing the process from the first manuscript to the final book, without which this book would be very much the poorer. I am also hugely appreciative of the skills of editor Julian Welch. I would also like to thank Sandra Buol who does so much to take my Australian writing to the world.

I would like to thank the Indigenous colleagues who have been so patient helping me move past the formal research to appreciate the nuances of the way knowledge systems are implemented within their cultures. In particular, David Kanosh, Margo Neale and Tyson Yunkaporta have been extraordinarily generous with their time and in sharing their unique insights. Although I have many associations with First Nations organisations in Australia, it was a new privilege to engage with the Sealaska Heritage Institute and through them explore so much about the Tlingit, Tsimshian and Haida cultures on the other side of the world.

Soon after writing her gorgeous contributions for this book, Jacqui Dark finally lost the battle with cancer at only 55. Despite her illness, she messaged me often, full of enthusiasm for the project. The world lost a wonderful opera singer and physics teacher, while I lost a very dear friend.

I was dependent on many people to help me understand their perspective, often very different from mine. A group I refer to as my Advisers were involved from the outset, spending copious amounts of time meeting and eating while exchanging ideas and experiences. Thank you most sincerely to Hilary Blackshaw, Tavish Bloom, Bridget Farmer, Julie King, Jeremy Meaden, Jane Rusden and Mary Thorpe. For seven years, I have been working with Paul Allen and Alice Steel; I have learnt so much from their insights.

There are so many people with whom I talked about the ideas within this book. I cannot mention them all, but I must particularly thank Avril Bowie, Xander Dark, Iain Davidson, Alison Edwards, Jen Ginsberg, Sam Ginsberg, Duane Hamacher, Tanya Hart, Marcus Houston, Stephanie Houston, Ian Irvine, Frida Jennings, Sue King-Smith, Win King-Smith, Lorraine LePlastrier, Shufen Lin, Shulan Lin, Richard Mayes, Meredith McKague, Sue McLeod, Joan Newman,

ACKNOWLEDGEMENTS

Jonathan Payne, David Reser, Jenny Rodger, Cheryl Ruan, Peggy Shaw, Michael Uniacke and Ash Vigus.

My family are the mainstay of my life. No one could ask for better than Rebecca, Rudi, Abigail and Leah Heitbaum, who not only support everything I do, but also keep me grounded.

But most of all, there is my husband, Damian Kelly. He has endured and encouraged my obsession with this book for the last three years. His expertise in IT, archaeology, education, librarianship, birding, photography and so much more has served me so well as a science writer. He researches, listens, suggests, calms, feeds and nurtures me. He is my partner in everything I do. Damian has earned the dedication in the front of this book many, many times over.

NOTES

Chapter 1: Knowledge makes us human

1 In academic reports, gene names are italicised while the names of disorders are not. So the neurofibromatosis type 1 gene would be written as *NF1* and the disorder as NF1. In books and websites for the general reader, it is common to avoid the italics, and for the author to make sure it is clear whether the reference is to the gene or the disorder. I shall follow the latter convention.

2 A. Alveshere, 'Forces of evolution', in *Explorations: An open invitation to biological anthropology*, American Anthropological Association, Arlington, 2019, p. 119.

3 There are many websites offering clearly explained information on NF1. They are remarkably consistent in what they describe, in both symptoms and cause. In Australia, the leading research centre working with NF1 is the Murdoch Children's Research Institute (see www.mcri.edu.au/impact/a-z-child-adolescent-health/m-n/neurofibromatosis). MCRI research on the cognitive impacts of NF1 is recognised globally and forms the basis for the understanding of NF1 as the knowledge gene. In the United States, the Children's Tumor Foundation's website is a wonderful resource (see www.ctf.org/understanding-nf/nf1). In the United Kingdom, see the Childhood Tumour Trust (www.childhoodtumourtrust.org.uk).

4 M.A. Summers, K.G. Quinlan, J.M. Payne, D.G. Little, K.N. North & A. Schindeler, 'Skeletal muscle and motor deficits in neurofibromatosis type 1', *Journal of Musculoskeletal and Neuronal Interactions*, vol. 15, no. 2, 2015, pp. 161–70.

5 Dr Vincent 'Vic' Riccardi has authored and co-authored many academic papers. One of his most significant is V.M. Riccardi, 'Historical background and introduction', in J.M. Friedman, D.H. Gutmann, M. Maccollin & V.M. Riccardi (eds), *Neurofibromatosis: Phenotype, natural history, and pathogenesis* (3rd edn), Johns Hopkins University Press, Baltimore, 1999, pp. 1–25. Vic has also worked with Professor Kathryn North AC, Director

379

of the Murdoch Children's Research Institute, and her colleagues for many decades. North's research is pertinent to understanding the cognitive impacts of NF1. Particularly relevant to understanding the role of NF1 as the knowledge gene are their joint ideas on the cognitive impacts of NF1. See for example, K.N. North, V. Riccardi, C. Samango-Sprouse, R. Ferner, B. Moore, E. Legius, N. Ratner & M.B. Denckla, 'Cognitive function and academic performance in neurofibromatosis. 1. Consensus statement from the NF1 Cognitive Disorders Task Force', *Neurology*, vol. 48, 1997, pp. 1121–27.

6 A large number of academic papers were used to summarise the cognitive implications of the NF1 disorder, and therefore what potentials might have been granted by the NF1 gene hundreds of thousands of years ago. In particular, Vic has been very much influenced by the research of Professor Kathryn North and her colleagues, of whom Jonathan Payne was particularly generous with his time and guidance. Consequently, many of the citations listed below represent the research Professor North has conducted with colleagues and are used as the primary source for the cognitive impact of the NF1 disorder. The specific data quoted here is from: K. North, P. Joy, D. Yuille, N. Cocks, E. Mobbs, P. Hutchins, K. McHugh & M. de Silva, 'Specific learning disability in children with neurofibromatosis type 1: Significance of MRI abnormalities', *Neurology*, vol. 44, no. 5, 1994, pp. 878–83; A.K. Chisholm, K.M. Haebich, N.A. Pride, K.S. Walsh, F. Lami, A. Ure, T. Maloof, A. Brignell, M. Rouel, Y. Granader, A. Maier, B. Barton, H. Darke, G. Dabscheck, V.A. Anderson, K. Williams, K.N. North & J.M. Payne, 'Delineating the autistic phenotype in children with neurofibromatosis type 1', *Molecular Autism*, vol. 13, no. 3, 2022, doi: 10.1186/s13229-021-00481-3.

7 B.C.L. Cota, J.G.M. Fonseca, L.O.C. Rodrigues, N.A. Rezende, P.B. Batista, V.M. Riccardi & L.M. Resende, 'Amusia and its electrophysiological correlates in neurofibromatosis type 1', *Archives of Neuropsychiatry*, vol. 76, no. 5, 2018, pp. 287–95, doi: 10.1590/0004-282X20180031.

8 J. Pfeifer & S. Hamann, 'The nature and nurture of congenital amusia: A twin case study', *Frontiers in Behavioral Neuroscience*, vol. 12, 2018, www.frontiersin.org/articles/10.3389/fnbeh.2018.00120.

9 Cota et al., 'Amusia and its electrophysiological correlates in neurofibromatosis type 1', p. 287.

10 Cota et al., 'Amusia and its electrophysiological correlates in neurofibromatosis type 1', p. 292.

11 I. Peretz & D. Vuvan, 'Prevalence of congenital amusia', *European Journal of Human Genetics*, vol. 25, 2017, p. 625.

12 For example, K. Szyfter & J. Wigowska-Sowińska, 'Congenital amusia: Pathology of musical disorder', *Journal of Applied Genetics*, vol. 63, 2022, pp. 127–31; Peretz & Vuvan, 'Prevalence of congenital amusia', pp. 625–30.

13 Peretz & Vuvan, 'Prevalence of congenital amusia', p. 625.

14 O. Sacks, *Musicophilia*, Picador, London, 2011, pp. 105–28.

15 S.L. Hyman, E.A. Shores & K.N. North, 'The nature and frequency of cognitive deficits in children with neurofibromatosis type 1', *Neurology*, vol. 65, 2005, pp. 1037–44.

16 C. Meneghetti, L. Miola, E. Toffalini, M. Pastore & F. Pazzaglia, 'Learning from navigation, and tasks assessing its accuracy: The role of visuospatial abilities and wayfinding inclinations', *Journal of Environmental Psychology*, vol. 75, 2021, doi: 10.1016/j.jenvp.2021.101614.

17 A.M. Clements-Stephens, S.L. Rimrodt, P. Gaur & L.E. Cutting, 'Visuospatial processing in children with neurofibromatosis type 1', *Neuropsychologia*, vol. 46, 2008, p. 691.

18 Hyman, Shores & North, 'The nature and frequency of cognitive deficits in children with neurofibromatosis type 1'.

19 B. Barton & K.N. North, 'Social skills of children with neurofibromatosis type 1', *Developmental Medicine & Child Neurology*, vol. 46, 2004, pp. 553–63; K.M. Haebich, D.P. Dao, N.A. Pride, B. Barton, K.S. Walsh, A. Maier, A.K. Chisholm, H. Darke, C. Catroppa, S. Malarbi, J.C. Wilkinson, V.A. Anderson, K.N. North & J.M. Payne, 'The mediating role of ADHD symptoms between executive function and social skills in children with neurofibromatosis type 1', *Child Neuropsychology*, vol. 28, no. 3, 2022, pp. 318–36.

20 A.K. Chisholm, F. Lami, K.M. Haebich, A. Ure, A. Brignell, T. Maloof, N.A. Pride, K.S. Walsh, A. Maier, M. Roue, Y. Granader, B. Barton, H. Darkel, I. Fuelscher, G. Dabscheck, V.A. Anderson, K. Williams, K.N. North & J.M. Payne, 'Sex- and age-related differences in autistic behaviours in children with neurofibromatosis type 1', *Journal of Autism and Developmental Disorders*, vol. 53, 2023, pp. 2835–50.

21 J.M. Payne, S.L. Hyman, E.A. Shores & K.N. North, 'Assessment of executive function and attention in children with neurofibromatosis type 1: Relationships between cognitive measures and real-world behavior', *Child Neuropsychology*, vol. 17, no. 4, 2011, pp. 313–29.

22 P. Batista, S. Lemos, L.O. Rodrigues & N. Rezende, 'Auditory temporal processing deficits and language disorders in patients with neurofibromatosis type 1', *Journal of Communication Disorders*, vol. 48, 2014, doi: 10.1016/j.jcomdis.2013.12.002.

23 S.S. Arnold, J.M. Payne, G. McArthur, K.N. North & B. Barton, 'Profiling the word reading abilities of school-age children with neurofibromatosis type 1', *Journal of the International Neuropsychological Society*, vol. 27, 2021, pp. 484–96.

Chapter 2: Illuminating an ancient knowledge gene

1 H. Chen, X. Lin, S. Lian & W. Zhu, 'Clinical manifestations and neuro-fibromatosis type 1 gene mutations of 25 patients with neurofibromatosis type 1 from 10 Chinese pedigrees', *Dermatologica Sinica*, vol. 38, no. 4, 2020, pp. 217–20.

2 G. Bloomfield, D. Traynor, S.P. Sander, D.M. Veltman, J.A. Pachebat & R.R. Kay, 'Neurofibromin controls macropinocytosis and phagocytosis in *Dictyostelium*', *eLife* 4:e04940, doi: 10.7554/eLife.04940.

3 M. Bergoug, M. Doudeau, F. Godin, C. Mosrin, B. Vallée & H. Bénédetti, 'Neurofibromin structure, functions and regulation', *Cells*, vol. 9, no. 11, 2365, doi: 10.3390/cells9112365.

4 G. Assum & C. Schmegner, 'NF1 gene evolution in mammals', in D. Kaufmann (ed.), *Neurofibromatoses*, Monographs in Human Genetics, Karger, Basel, 2008, vol. 16, pp. 103–12; B. Bartelt-Kirbach & D. Kaufmann, 'Insights into NF1 from evolution', in M. Upadhyaya & D.N. Cooper (eds), *Neurofibromatosis Type 1 Molecular and Cellular Biology*, Springer, Heidelberg, 2012, pp. 253–68.

5 The federal agency, the National Institutes of Health (NIH), describes the GenBank website (www.ncbi.nlm.nih.gov/genbank) as a genetic sequence database, an annotated collection of all publicly available DNA sequences.

6 The 1000 Genomes Project database (www.internationalgenome.org) is also freely available. For those who are technically minded and genetically adept, Andrea reported that the human NF1 allele varies from chimps at GRCh37 17:29677329, which is the first base in the codon. No codon pro-ducing an (ancestral, chimp-like) T2484 neurofibromin residue has ever been documented among modern humans, aside from somatic mutations in tumour tissue. Therefore, the human wild type NF1 allele is derived compared to all other primates and is fixed in the modern human popula-tion: A.J. Alveshere & V.M. Riccardi, 'The human NF1 locus: A unique hominin genotype with potential evolutionary significance', Paper pre-sented at the 89th Annual Meeting of the American Association of Physical Anthropologists, 18 April 2020, Los Angeles, CA.

7 S. Almécija, A.S. Hammond, N.E. Thompson, K.D. Pugh, S. Moyà-Solà & D.M. Alba, 'Fossil apes and human evolution', *Science*, vol. 372, no. 6542, eabb4363, doi: 10.1126/science.abb4363.

8 According to the Australian Museum, the most recent definitions of *hominid* and *hominin* are as follows: Hominid—the group consisting of all modern and extinct Great Apes (that is, modern humans, chimpanzees, gorillas and orangutans, plus all their immediate ancestors). Hominin—the group consisting of modern humans, extinct human species and all our immediate ancestors (including members of the genera *Homo*, *Australopithecus*, *Paranthropus* and *Ardipithecus*). See B. Blaxland, 'Hominid and hominin— what's the difference?', 2 October 2020, https://australian.museum/learn/science/human-evolution/hominid-and-hominin-whats-the-difference.

9 The Max Planck Institute for Evolutionary Anthropology genome projects can be viewed on their website for no cost: www.eva.mpg.de/genetics/genome-projects. The genomes are free for anyone to search.

10 R. Teague & R. McRae, 'Ancient DNA and Neanderthals', Smithsonian National Museum of Natural History, https://humanorigins.si.edu/evidence/genetics/ancient-dna-and-neanderthals.

11 Alveshere, 'Forces of evolution', in *Explorations: An open invitation to biological anthropology*, p. 129.

12 K. Esoh & A. Ambroise Wonkam, 'Evolutionary history of sickle-cell mutation: Implications for global genetic medicine', *Human Molecular Genetics*, vol. 30, no. R1, 2021, pp. R119–R128, doi: 10.1093/hmg/ddab004.

13 P. Charlier, N. Benmoussa, P. Froesch, I. Huynh-Charlier & A. Balzeau, 'Did Cro-Magnon 1 have neurofibromatosis type 1?', *The Lancet*, vol. 391, no. 10127, 2018, p. 1259.

14 V.M. Riccardi, 'NF1 and the Praxitype', *JSM Genetics and Genomics*, vol. 2, no. 1, 2015, p. 1006.

15 W. Enard, M. Przeworski, S.E. Fisher, C.S. Lai, V. Wiebe, T. Kitano, A.P. Monaco & S. Pääbo, 'Molecular evolution of FOXP2, a gene involved in speech and language', *Nature*, no. 418, 2002, pp. 869–72; J. Zhang, D.M. Webb & O. Podlaha, 'Accelerated protein evolution and origins of human-specific features: FOXP2 as an example', *Genetics*, no. 162, 2002, pp. 1825–35.

Chapter 3: A fresh look at Indigenous memory systems

1 L. Kelly, *Knowledge and Power in Prehistoric Societies: Orality, memory and the transmission of culture*, Cambridge University Press, New York, 2015, pp. 127–28.

2 R.H. Kaschula, 'Imbongi and Griot: Toward a comparative analysis of oral poetics in Southern and West Africa', *Journal of African Cultural Studies*, vol. 12, no. 1, 1999, pp. 55–76.

3 P. Hage, 'Speculations on Puluwatese mnemonic structure', *Oceania*, vol. 49, no. 2, 1978, pp. 81–95.

4 J. Vansina, *Oral Tradition as History*, University of Wisconsin Press, Madison, 1985, p. 182.

5 Phillips Indian Educators, 'Jim Rock', n.d., http://pieducators.com/wisdom/jim_rock.

6 L.C. Wyman & F.L. Bailey, *Navaho Indian Ethnoentomology*, The University of New Mexico Press, Albuquerque, 1964.

7 'Primary orality' came to prominence as an academic term through the work of Walter Ong in *Orality and Literacy: The technologizing of the word* (1982) and Ruth Finnegan in *Literacy and Orality: Studies in the technology of communication* (1988).

8 T.G.H. Strehlow, *Songs of Central Australia*, Angus & Robertson, Sydney, 1971, p. 257.

9 Parks Australia, 'Tjukurpa', n.d., https://parksaustralia.gov.au/uluru/discover/culture/tjukurpa.

10 Vansina, *Oral Tradition as History*, pp. 19–21.

11 A. Ortiz, *The Tewa World: Space, time, being, and becoming in a Pueblo society*, University of Chicago Press, Chicago, 1969, pp. xviii, 13–15. This is really illuminating when read alongside R.I. Ford, 'The color of survival', *Discovery*, 1980, pp. 16–29.

12 J. MacDonald, *The Arctic Sky: Inuit astronomy, star lore, and legend*, Royal Ontario Museum/Nunavut Research Institute, Ontario, 1998, p. 168.

13 C. Goddard & A. Kalotas, *Punu: Yankunytjatjara plant use: Traditional methods of preparing foods, medicines, utensils and weapons from native plants*, North Ryde, Angus & Robertson, 2002, p. iv.

14 Margo was my co-author on *Songlines: The power and promise*, Thames & Hudson, Melbourne, 2020.

15 E. McDinny, quoted in J. Bradley, *Singing Saltwater Country: Journey to the songlines of Carpentaria*, Allen & Unwin, Crows Nest, 2010, p. 29.

16 R.M.W. Dixon & G. Koch, *Dyirbal Song Poetry: The oral literature of an Australian rainforest people*, University of Queensland Press, St Lucia, 1996.

17 P.R. Schmidt, *Historical Archaeology in Africa: Representation, social memory, and oral traditions*, AltaMira Press, Lanham, 2006.

18 Bradley, *Singing Saltwater Country*, p. 216.

19 L.D. Minc, 'Scarcity and survival: The role of oral tradition in mediating subsistence crises', *Journal of Anthropological Archaeology*, vol. 5, no. 1, 1986, pp. 39–113.

20 P.D. Nunn, *The Edge of Memory*, Bloomsbury, London, 2018, in particular pp. 63–107. See also P.D. Nunn, *Worlds in Shadow: Submerged lands in science, memory and myth*, Bloomsbury, London, 2021.

21 D. Hamacher, *The First Astronomers: How Indigenous elders read the stars*, Allen & Unwin, Crows Nest, 2022.

22 R. Nair, 'Archeological find affirms Heiltsuk Nation's oral history', *CBC News*, 31 March 2017, www.cbc.ca/news/canada/british-columbia/archeological-find-affirms-heiltsuk-nation-s-oral-history-1.4046088.

23 D.A. Fletcher & J.A. Theriot, 'An introduction to cell motility for the physical scientist', *Physical Biology*, vol. 1, no 1, 2004, pp. T1–10.

Chapter 4: Why oral cultures mark their landscapes

1 Unfortunately, given the scope of this book, the actual mnemonic technologies can only be mentioned briefly. They are dealt with in more detail in my book *Knowledge and Power in Prehistoric Societies* (2015). These topics are also covered in *The Memory Code* (2016), while how to implement the methods yourself is described in detail in *Memory Craft* (2019). A full bibliography of well over 800 research items can be found on my website, www.lynnekelly.com.au, under 'Memory Systems Research'.

2 The songlines of the Yanyuwa people are detailed in John Bradley's fascinating book written with the Yanyuwa, *Singing Saltwater Country* (2010).

3 I work with many Indigenous colleagues, but hugely influential in my understanding of songlines is Margo Ngawa Neale, Head of the Centre for Indigenous Knowledges at the National Museum of Australia and my co-author on *Songlines: The power and promise* and *Songlines for Younger Readers*. Tyson Yunkaporta (Apalech Clan, Far North Queensland) is Senior Lecturer in Indigenous Knowledges at Deakin University and the author of *Sand Talk* (2019). Tyson and I are part of a team from four universities researching the differences between the method of *loci* and Aboriginal songlines.

4 A full description of my method for memorising Chinese is available on my website, www.lynnekelly.com.au.

5 A.S. Lee, J. Rock & C. O'Rourke, *Dakota/Lakota Star Map Constellation Guidebook: An introduction to D(L)akota star knowledge*, Native Skywatchers, Minnesota, 2014.

6 K. Monteleone, A.E. Thompson & K.M. Prufer, 'Virtual cultural landscapes: Geospatial visualizations of past environments', *Archaeological Prospection*, vol. 28, no. 3, 2021, doi: 10.1002/arp.1830, pp. 379–401. Specific data is quoted in a lecture by Dr Monteleone for Sealaska Heritage Institute, www.youtube.com/watch?v=a6kuJNHi0V4.

7 See www.sealaskaheritage.org. Sealaska Heritage Institute is a private non-profit founded in 1980 to perpetuate and enhance Tlingit, Haida and Tsimshian cultures of south-eastern Alaska.

8 Dr Thomas Thornton presents a particularly interesting lecture that should be watched when considering David Kanosh's description of learning the flood stories from his grandparents: 'Alpine cairns and social and environmental change in SE Alaska', www.youtube.com/watch?v=ncTQ10-oLtU. See also: W.J. Hunt Jr, R.J. Hartley, B. McCune, N. Ali & T.F. Thornton, 'Maritime alpine cairns in Southeast Alaska: A multidisciplinary exploratory study', *Anthropology Faculty Publications*, vol. 129, 2016, http://digital commons.unl.edu/anthropologyfacpub/129.

9 P.S.C. Taçon, S. Thompson, K. Greenwood, A. Jalandoni, M. Williams & M. Kottermair, 'Marra Wonga: Archaeological and contemporary First Nations interpretations of one of central Queensland's largest rock art sites', *Australian Archaeology*, vol. 88, no. 2, 2022, pp. 159–79.

10 V.E. Garfield & L.A. Forrest, *The Wolf and the Raven: Totem poles of southeastern Alaska*, University of Washington Press, Seattle, 1961.

11 David Kanosh's performance at the Sealaska Heritage Celebration 2018 with the Tlingit dance group Sitka Kaagwaantaan Dancers can be viewed at www.youtube.com/watch?v=oB_yJ3I6fxk. I highly recommend all the videos and lectures being produced by the Sealaska Heritage Institute.

Chapter 5: The oral knowledges skill set

1 D. Taylor, G. Dezecache & M. Davila-Ross, 'Affective prosody in grunts of young chimpanzees', *Revue de primatologie*, vol. 13, 2022, http://journals.openedition.org/primatologie/14376.

2 I. Morley, 'Hunter-gatherer music and its implications for identifying intentionality in the use of acoustic space', in C. Scarre & G. Lawson (eds), *Archaeoacoustics*, McDonald Institute for Archaeological Research, Cambridge, 2006, p. 103.

3 R.H. Finnegan, *Oral Poetry: Its nature, significance and social context*, Cambridge University Press, Cambridge, 1977, pp. 98, 119–21; W.J. Ong, 'African talking drums and oral noetics', *New Literary History*, vol. 8, no. 3, 1977, pp. 411–29.

4 Two excellent resources to watch Australian Aboriginal dances are: P. Cameron (writer), *Dance on Your Land*, Woomera Aboriginal Corporation, Ronin Films, Civic Square, 1993; and T. Graham (writer), *Ceremony: The Djungguwan of Northeast Arnhem Land*, Film Australia, Lindfield, 2006.

5 Kelly, *Knowledge and Power in Prehistoric Societies*, pp. xv–xviii.

6 D.C. Rubin, *Memory in Oral Traditions: The cognitive psychology of epic, ballads, and counting-out rhymes*, Oxford University Press, New York, 1995, p. 112.

7 H. Morphy, *Aboriginal Art*, Phaidon Press, London, 1998, pp. 183–84.

8 M.J. King, D.P. Katz, L.A. Thompson & B.N. Macnamara, 'Genetic and environmental influences on spatial reasoning: A meta-analysis of twin studies', *Intelligence*, vol. 73, 2019, pp. 65–77.

9 M. Neale & L. Kelly, *Songlines: The power and promise*, Thames & Hudson, Melbourne, 2020. There are very many references to Aboriginal knowledge being linked to songlines, but it is not only that of men. Women's knowledge is often neglected. A fascinating work is C.J. Ellis & L. Barwick, 'Antikirinja women's song knowledge 1963–72: Its significance in Antikirinja culture', in P. Brock (ed.), *Women, Rites and Sites: Aboriginal women's cultural knowledge*, Allen & Unwin, Crows Nest, 1989, pp. 21–40.

10 D.L. Fixico, *The American Indian Mind in a Linear World: American Indian studies and traditional knowledge*, Routledge, New York, 2003, p. 25.

11 D. Turnbull, 'Maps, narratives and trails: Performativity, hodology and distributed knowledges in complex adaptive systems—an approach to emergent mapping', *Geographical Research*, vol. 45, no. 2, 2007, p. 142.

12 M. Campbell, 'Memory and monumentality in the Rarotongan landscape', *Antiquity*, vol. 80, no. 307, 2006, pp. 102–17.

13 J. Vansina, *Oral Tradition as History*, University of Wisconsin Press, Madison, 1985, p. 45; E.M. McClelland, *The Cult of Ifa among the Yoruba*, Ethnographica, London, 1982, p. 71.

14 Tyson's words here come from a document he wrote for a research team in which we are both involved. He gave permission for me to quote him here.

15 There are many resources describing the extraordinary astronomical skills of First Nations cultures, all including the way the knowledge keepers decode the spatial knowledge from song. Four of my favourites on the topic are: R.D. Haynes, 'Astronomy and the Dreaming: The astronomy of the Aboriginal Australians', in H. Selin (ed.), *Astronomy Across Cultures: The history of non-Western astronomy*, Kluwer Academic Publishers, Dordrecht, 2000; J. MacDonald, *The Arctic Sky: Inuit astronomy, star lore, and legend*, Royal Ontario Museum/Nunavut Research Institute, Ontario, 1998; J.E. Reyman, 'Priests, power, and politics: Some implications of socio-ceremonial control', in J.B. Carlson & J.W. Judge (eds), *Astronomy and Ceremony in the Prehistoric Southwest*, Maxwell Museum of Anthropology, Albuquerque, 1987, pp. 121–47; and D. Hamacher, *The First Astronomers: How Indigenous elders read the stars*, Allen & Unwin, Crows Nest, 2022.

16 J. Green, C. Algy & F. Meakins, 'Tradition and innovation: How we are documenting sign language in a Gurindji community in northern Australia', *The Conversation*, 12 December 2022, https://the conversation.com/tradition-and-innovation-how-we-are-documenting-sign-language-in-a-gurindji-community-in-northern-australia-194524?.

17 Kelly, *Knowledge and Power in Prehistoric Societies*, pp. 72–91; Neale & Kelly, *Songlines*, pp. 153–77.

18 H. Morphy, *Ancestral Connections: Art and an Aboriginal system of knowledge*, University of Chicago Press, Chicago, 1991, p. 167.

19 C. Lévi-Strauss, *The Savage Mind*, University of Chicago Press, Chicago, 1966, p. 64.

20 C.J. Couch, 'Oral technologies: A cornerstone of ancient civilizations?', *The Sociological Quarterly*, vol. 30, no. 4, 1989, p. 589. It was Carl Couch's paper that gave me the confidence to move into archaeology. He worked with Professor Sarina Chen at the University of Northern Iowa, whom I contacted for help in those early days. She could not have been more encouraging, even inviting me to present my first conference paper, 'Stonehenge: A monument to orality', at the National Communication Association in Chicago in 2009.

21 C.J. Couch, *Constructing Civilizations*, JAI Press, Greenwich, 1990; C.J. Couch, *Information Technologies and Social Orders*, Aldine de Gruyter, New York, 1996; C.J. Couch & S.-L. Chen, 'Orality, literacy and social structure', in D.R. Maines & C.J. Couch (eds), *Communication and Social Structure*, Charles C. Thomas, Springfield, 1988.

22 J. Fentress & C. Wickham, *Social Memory*, Blackwell, Oxford, 1992, p. 50.

23 W.P. Murphy, 'Secret knowledge as property and power in Kpelle society: Elders versus youth', *Africa: Journal of the International African Institute*, vol. 50, no. 2, 1980, p. 199.

24 Lévi-Strauss, *The Savage Mind*, p. 6.

25 C.B. Keibel, 'Memory sticks and other mnemonic devices', *The Nigerian Field*, vol. 55, no. 3/4, 1990, pp. 91–98.

26 W.R. Bascom, *Sixteen Cowries: Yoruba divination from Africa to the New World*, Indiana University Press, Bloomington, 1980.

27 McClelland, *The Cult of Ifa among the Yoruba*.

28 B. Tedlock, *Time and the Highland Maya* (rev. edn), University of New Mexico Press, Albuquerque, 1992, pp. 163–67.

29 A.J. Alveshere, L.S. Kelly & V.M. Riccardi, 'Art, orality, and migration: The roles of NF1, mnemonics, and somatic adaptation in the hominin bio cultural toolkit', Paper presented at the 91st Annual Meeting of the American Association of Biological Anthropologists, 29 March 2022, Denver, CO.

30 František Kratochvíl, in 2009, was a linguist at my university, La Trobe. He is now a professor at Palacky University Olomouc in the Czech Republic. His response to my question was sent in an email, 14 October 2009.

31 More information can be found on the Williams Syndrome Association website, https://williams-syndrome.org/resources/factinfo-sheet/music-

and-williams-syndrome. Amusia research is reported in M. Lense, C. Shivers & E. Dykens, '(A)musicality in Williams syndrome: Examining relationships among auditory perception, musical skill, and emotional responsiveness to music', *Frontiers in Psychology*, vol. 4, 2013, doi: 10.3389/fpsyg.2013.00525. For spatial impacts, see B. Landau & K. Ferrara, 'Space and language in Williams syndrome: Insights from typical development', *Wiley Interdisciplinary Reviews: Cognitive Science*, 2013, doi: 10.1002/wcs.1258.

32 Tyson's words here come from a document he wrote for a research team in which we are both involved. He gave permission for me to quote him here.

Chapter 6: Questions about neurodiversity

1 S. Petru, 'Identity and fear—burials in the Upper Palaeolithic', *Documenta Praehistorica*, vol. 45, 2019, pp. 6–13, doi: 10.4312/dp.45.1.

2 E. Trinkaus, 'An abundance of developmental anomalies and abnormalities in Pleistocene people', *PNAS*, vol. 115, no. 47, 2018, p. 11941.

3 L. Surugue, 'Why this Paleolithic burial site is so strange (and so important)', 22 February 2018, www.sapiens.org/archaeology/paleolithic-burial-sunghir.

4 Hugely influential on my thinking was Thomas Armstrong, *The Power of Neurodiversity: Unleashing the advantages of your differently wired brain*, Da Capo, Cambridge, 2010.

5 From the Australian *National Guideline for the Assessment and Diagnosis of Autism Spectrum Disorders*, www.autismcrc.com.au/access/national-guideline.

6 Penny Spikins is Senior Lecturer in the Archaeology of Human Origins at the University of York and writes extensively on the evidence for autism and human origins. For an easy entry to her work, see P. Spikins, 'How our autistic ancestors played an important role in human evolution', *The Conversation*, 28 March 2017, https://theconversation.com/how-our-autistic-ancestors-played-an-important-role-in-human-evolution-73477. See also P. Spikins & B. Wright, *The Prehistory of Autism*, Rounded Globe, 2016. Also excellent is P. Spikins, B. Wright & D. Hodgson, 'Are there alternative adaptive strategies to human pro-sociality? The role of collaborative morality in the emergence of personality variation and autistic traits', *Time and Mind*, vol. 9, no. 4, 2016, pp. 289–313.

7 I was also heavily influenced by S. Silberman, *Neurotribes: The legacy of autism and how to think smarter about people who think differently*, Allen & Unwin, Crows Nest, 2015 and T. Grandin & R. Panek, *The Autistic Brain: Exploring the strength of a different kind of mind*, Rider Books, London, 2014.

8 F. Happé & U. Frith, 'The weak coherence account: Detail-focused cognitive style in autism spectrum disorders', *Journal of Autism and Developmental Disorders*, vol. 36, no. 1, 2006, pp. 5–25.

9   Australasian ADHD Professionals Association (AADPA), https://adhdguide
    line.aadpa.com.au/lived-experience-factsheet.

10 Thomas Armstrong's *The Power of Neurodiversity* devotes an entire chapter to
    ADHD.

11 Dyslexia theory draws on a range of resources, particularly Armstrong,
    The Power of Neurodiversity; H.T. Vestergaard & M. David, 'Developmental
    dyslexia: Disorder or specialization in exploration?', *Frontiers in Psychology*,
    vol. 13, 2022; C. von Károlyi, E. Winner, W. Gray & G.F. Sherman,
    'Dyslexia linked to talent: Global visual-spatial ability', *Brain and Language*,
    vol. 85, no. 3, 2003, pp. 427–31.

12 C. Doust, P. Fontanillas, E. Eising, S.D. Gordon, Z. Wang, G. Alagöz,
    B. Molz, 23andMe Research Team, Quantitative Trait Working Group
    of the GenLang Consortium, B.S. Pourcain, C. Francks, R.E. Marioni,
    J. Zhao, S. Paracchini, J.B. Talcott, A.P. Monaco, J.F. Stein, J.R. Gruen,
    R.K. Olson, E.G. Willcutt, J.C. DeFries, B.F. Pennington, S.D. Smith,
    M.J. Wright, N.G. Martin, A. Auton, T.C. Bates, S.E. Fisher &
    M. Luciano, 'Discovery of 42 genome-wide significant loci associated with
    dyslexia', *Nature Genetics*, vol. 54, no. 11, 2022, pp. 1621–29; C. Doust,
    P. Fontanillas, E. Eising, S.D. Gordon, Z. Wang, G. Alagöz, B. Molz,
    23andMe Research Team, Quantitative Trait Working Group of the
    GenLang Consortium, B.S. Pourcain, C. Francks, R.E. Marioni, J. Zhao,
    S. Paracchini, J.B. Talcott, A.P. Monaco, J.F. Stein, J.R. Gruen, R.K. Olson,
    E.G. Willcutt, J.C. DeFries, B.F. Pennington, S.D. Smith, M.J. Wright,
    N.G. Martin, A. Auton, T.C. Bates, S.E. Fisher & M. Luciano, 'Author
    Correction: Discovery of 42 genome-wide significant loci associated with
    dyslexia', *Nature Genetics*, vol. 55, no. 3, 2023, p. 520.

13 T. Simonite, 'Colour blindness may have hidden advantages', *Nature*,
    2005, doi: 10.1038/news051205-1; J. Bosten, J.D. Robinson, G. Jordan &
    J.D. Mollon, 'Multidimensional scaling reveals a color dimension unique to
    "color-deficient" observers', *Current Biology*, vol. 15, 2005, pp. R950–R952.

14 C.J. Dance, M. Jaquiery, D.M. Eagleman, D. Porteous, A. Zeman &
    J. Simner, 'What is the relationship between aphantasia, synaesthesia
    and autism?', *Consciousness and Cognition*, vol. 89, 2021, doi: 10.1016/
    j.concog.2021.103087.

15 D. Cavedon-Taylor, 'Aphantasia and psychological disorder: Current con-
    nections, defining the imagery deficit and future directions', *Frontiers in
    Psychology*, vol. 13, 2022, doi: 10.3389/fpsyg.2022.822989.

16 S.G. Miller, 'Why other senses may be heightened in blind people',
    Live Science, 23 March 2017, www.livescience.com/58373-blindness-
    heightened-senses.html.

17 M.V. Gonzaga, 'Blindness—Evolutionary regression? Maybe not!', *Biology Online*, 28 July 2021, www.biologyonline.com/articles/blindness-evolutionary-regression-maybe-not.

18 H. Hamilton, 'Memory skills of deaf learners: Implications and applications', *American Annals of the Deaf*, vol. 156, no. 4, 2011, pp. 402–23.

19 Michael Uniacke is a deaf activist and an author who writes extensively about deafness. He quotes geneticist Rachel Burt in his 2015 book *Deafness Down*, to which the sequel is *Deafness Gain*. Both were published by The Unguarded Quarter in Castlemaine, Victoria (see www.tuq.pub/about-michael-uniacke).

Chapter 7: The prehistory of music through a knowledge gene lens

1 Cota et al., 'Amusia and its electrophysiological correlates in neurofibromatosis type 1'.

2 I first came across Iain Morley's work through I. Morley, 'Hunter-gatherer music and its implications for identifying intentionality in the use of acoustic space', in C. Scarre & G. Lawson (eds), *Archaeoacoustics*, McDonald Institute for Archaeological Research, Cambridge, 2006, pp. 95–105.

3 I. Morley, *The Prehistory of Music: Human evolution, archaeology, and the origins of musicality*, Oxford University Press, Oxford, 2013.

4 Morley, *The Prehistory of Music*, pp. 6–7, 308.

5 J. Powell, *How Music Works: A listener's guide to harmony, keys, chords, perfect pitch and other secrets of a good tune*, Little, Brown & Company, Boston, 2011, p. 124.

6 S.E. Freidline, K.E. Westaway, R. Joannes-Boyau, P. Duringer, J.L. Ponche, M.W. Morley, V.C. Hernandez, M.S. McAllister-Hayward, H. McColl, C. Zanolli, P. Gunz, I. Bergmann, P. Sichanthongtip, D. Sihanam, S. Boualaphane, T. Luangkhoth, V. Souksavatdy, A. Dosseto, Q. Boesch, E. Patole-Edoumba, F. Aubaile, F. Crozier, E. Suzzoni, S. Frangeul, N. Bourgon, A. Zachwieja, T.E. Dunn, A.M. Bacon, J.J. Hublin, L. Shackelford & F. Demeter, 'Early presence of *Homo sapiens* in Southeast Asia by 86-68 kyr at Tam Pà Ling, Northern Laos', *Nature Communications*, vol. 14, no. 1, 2023, p. 3193.

7 C. Clarkson, Z. Jacobs, B. Marwick, R. Fullagar, L. Wallis, M. Smith, R.G. Roberts, E. Hayes, K. Lowe, X. Carah, S.A. Florin, J. McNeil, D. Cox, L.J. Arnold, Q. Hua, J. Huntley, H.E.A. Brand, T. Manne, A. Fairbairn, J. Shulmeister, L. Lyle, M. Salinas, M. Page, K. Connell, G. Park, K. Norman, T. Murphy & C. Pardoe, 'Human occupation of northern Australia by 65,000 years ago', *Nature*, vol. 547, 2017, pp. 306–10.

8 M. Spitzer, *The Musical Human: A history of life on Earth*, Bloomsbury Publishing, London, 2021, pp. 281–341.

9 R.J. Jao Keehn, J.R. Iversen, I. Schulz & A.D. Patel, 'Spontaneity and diversity of movement to music are not uniquely human', *Current Biology*, vol. 29, no. 13, 2019, pp. R621–R622.

10 S. Mithen, *The Singing Neanderthals*, Harvard University Press, Cambridge, 2006, pp. 118–21.

11 A. Harvey, *Music, Evolution, and the Harmony of Souls*, Oxford University Press, Oxford, 2017; see especially pp. 245 and 253.

12 Morley, *The Prehistory of Music*, pp. 201–05.

13 C.R. Darwin, *The Descent of Man, and Selection in Relation to Sex*, John Murray, London, 1871, p. 365.

14 S. Pinker, *How the Mind Works*, W.W. Norton & Company, New York, 2009, p. 534.

15 Sacks, *Musicophilia*, pp. 263–4.

16 Harvey, *Music, Evolution, and the Harmony of Souls*, pp. 97–169.

17 Morley, *The Prehistory of Music*, p. 281.

18 Sacks, *Musicophilia*, p. 266.

19 Sacks, *Musicophilia*, p. 372–80.

20 NHS Chorus-19, 'Somewhere Over the Rainbow', www.youtube.com/watch?v=oxYRaUCXrXg.

21 A.D. Patel, 'Musical rhythm, linguistic rhythm, and human evolution', *Music Perception*, vol. 24, no. 1, 2006, pp. 99–104.

22 Mithen, *The Singing Neanderthals*, pp. 46, 59–61.

23 Mithen, *The Singing Neanderthals*, pp. 53–54.

24 Y.T. Tan, G.E. McPherson, I. Peretz, S.F. Berkovic & S.J. Wilson, 'The genetic basis of music ability', *Frontiers in Psychology*, vol. 5, 2014, pp. 1–19.

25 I. Peretz, J. Ross, C.V. Bourassa, L.-P.L. Perreault, P.A. Dion, M.W. Weiss, M. Felezeu, G.A. Rouleau & M.-P. Dubé, 'Do variants in the coding regions of FOXP2, a gene implicated in speech disorder, confer a risk for congenital amusia?', *Annals of the New York Academy of Sciences*, vol. 1517, no. 1, 2022, pp. 279–85.

26 Sacks, *Musicophilia*, p. 260.

27 Sacks, *Musicophilia*, p. 262.

28 Powell, *How Music Works*, p. 200.

29 See 'Tradition: Scottish women sing so beautifully as they waulk expensive Harris Tweed', 1941, www.youtube.com/watch?v=3A-JbxqiQyw.

30 H. van Praag, 'Neurogenesis and exercise: Past and future directions', *Neuromolecular Medicine*, vol. 10, no. 2, 2008, pp. 128–40.

31 Morley, *The Prehistory of Music*, pp. 29–31.

32 Morley, *The Prehistory of Music*, p. 17.

33 Morley, *The Prehistory of Music*, p. 38.

34 Spitzer, *The Musical Human*, p. 142.

35 Morley, *The Prehistory of Music*, p. 126.

36 Morley, *The Prehistory of Music*, pp. 32–98.

37 G. Lawson & F. d'Errico, 'Microscopic, experimental and theoretical re-assessment of Upper Paleolithic bird-bone pipes from Isturitz, France: Ergonomics of design, systems of notation and the origins of musical tradition', in E. Hickmann & R. Eichmann (eds), *The Archaeology of Sound: Origin and organization*, Papers from the 2nd Symposium of the Study Group on Music Archaeology, *Orient-Archäologie, 12*, 2002, pp. 119–142.

38 F. d'Errico, 'Palaeolithic origins of artificial memory systems: An evolutionary perspective', in C. Renfrew & C.E. Scarre (eds), *Cognition and Material Culture: The archaeology of symbolic storage*, McDonald Institute for Archaeological Research, Cambridge, 1998, pp. 19–50.

39 Kelly, *Knowledge and Power in Prehistoric Societies*, p. 77.

40 Morley, *The Prehistory of Music*, pp. 100–09.

41 L. Backwell, J. Bradfield, K. Carlson, T. Jashashvili, L. Wadley & F. d'Errico, 'The antiquity of bow-and-arrow technology: Evidence from Middle Stone Age layers at Sibudu cave', *Antiquity*, vol. 92, no. 362, 2018, pp. 289–303.

42 L. Metz, J.E. Lewis & L. Slimak, 'Bow-and-arrow, technology of the first modern humans in Europe 54,000 years ago at Mandrin, France', *Science Advances*, vol. 9, no. 8, 2023, doi: 10.1126/sciadv.add4675.

43 Morley, *The Prehistory of Music*, p. 121.

44 J. Shepherd & P. Wicke, *Continuum Encyclopedia of Popular Music of the World: Volume II: Performance and production*, Bloomsbury Publishing, 2003, p. 276.

45 A. Kleinloog, 'Rocks that gong in the Midlands of Kwazulu-Natal', *The Digging Stick*, vol. 36, no. 3, 2019, pp. 1–6.

46 Morley, *The Prehistory of Music*, p. 119.

47 Morley, *The Prehistory of Music*, pp. 114–15; A. Kossykh, 'Music and sounds in ancient Europe', in S. de Angeli, A.A. Both, S. Hagel, P. Holmes, R.J. Pasalodos & C.S. Lund (eds), *Contributions from the European Music Archaeology Project*, EMAP, 2018, pp. 34–9.

48 P.G. Bahn & J. Vertut, *Journey through the Ice Age*, Weidenfeld & Nicolson, London, 1997, p. 113.

49 Y. Garfinkel, *Dancing at the Dawn of Agriculture*, University of Texas Press, 2003.

50 B. Fazenda, C. Scarre, R. Till, R.J. Pasalodos, M.R. Guerra, C. Tejedor, R.O. Peredo, A. Watson, S. Wyatt, C.G. Benito, H. Drinkall & F. Foulds, 'Cave acoustics in prehistory: Exploring the association of Palaeolithic visual motifs and acoustic response', *The Journal of the Acoustical Society of America*, vol. 142, no. 3, 2017, p. 1332.

51 I. Reznikoff & M. Dauvois, 'La dimension sonore des grottes ornées', *Bulletin de la Société préhistorique française*, vol. 85, no. 8, 1988, 238–46; Morley, *The Prehistory of Music*, pp. 115–19.

52 Morley, *The Prehistory of Music*, p. 119.

53 Sacks, *Musicophilia*, pp. 372, 380–85.

54 S. Walsh, R. Causer & C. Brayne, 'Does playing a musical instrument reduce the incidence of cognitive impairment and dementia? A systematic review and meta-analysis', *Aging & Mental Health*, vol. 25, no. 4, 2021, pp. 593–601.

Chapter 8: Prehistoric art and the knowledge gene

1 Kelly, *Knowledge and Power in Prehistoric Societies*, p. 237.

2 P.G. Bahn, *Images of the Ice Age*, Oxford University Press, Oxford, 2016, pp. 33–63.

3 R. Kellogg, *Analyzing Children's Art*, Mayfield, Palo Alto, 1970.

4 G.M. Morriss-Kay, 'The evolution of human artistic creativity', *Journal of Anatomy*, vol. 216, no. 2, 2010, pp. 158–76.

5 I. Davidson, 'Symbolism and becoming a hunter-gatherer', Paper presented at the IFRAO Congress: Pleistocene art of the world, 2010, Ariège, Pyrenees, France, p. 3.

6 I cannot recommend the works by Polly Nooter Roberts and Allen Roberts on Luba art more highly, even if just to admire the aesthetics. But there is so much more in these exquisite artworks. See M. Nooter Roberts & A.F. Roberts (eds), *Memory: Luba art and the making of history*, Museum for African Art, New York, 1996; M. Nooter Roberts & A.F. Roberts, 'Memory: Luba art and the making of history', *African Arts*, vol. 29, no. 1, 1996, pp. 23–103; and M. Nooter Roberts & A.F. Roberts, *Luba*, 5 Continents Editions, Milan, 2007.

7 'Dialogue: Rebeca Méndez and Polly Nooter Roberts', UCLA School of the Arts and Architecture, 31 May 2018, www.arts.ucla.edu/single/dialogue-rebeca-mendez-and-polly-nooter-roberts.

8 L. Rademaker, G. Maralngurra, J. Goldhahn, K. Mangiru, P. Tacon & S. May, 'Friday essay: "This is our library"—How to read the amazing archive of First Nations stories written on rock', *The Conversation*, 25 March 2022, https://theconversation.com/friday-essay-this-is-our-library-how-to-read-the-amazing-archive-of-first-nations-stories-written-on-rock-176886.

9 Cambridge Dictionary, https://dictionary.cambridge.org/dictionary/english/utilitarian.

10 J. Joordens, F. d'Errico, F. Wesselingh, S. Munro, J. de Vos, J. Wallinga, C. Ankjærgaard, T. Reimann, J.R. Wijbrans, K.F. Kuiper, H.J. Mücher,

H. Coqueugniot, V. Prié, I. Joosten, B. van Os, A.S. Schulp, M. Panuel, V. van der Haas, W. Lustenhouwer, J.J.G. Reijmer & W. Roebroeks, '*Homo erectus* at Trinil on Java used shells for tool production and engraving', *Nature*, vol. 518, 2015, pp. 228–31.

11 Bahn, *Images of the Ice Age*, p. 23.

12 R. Baxter, '125,000-year-old bone engravings may be oldest ever example of Denisovan art', *IFL Science*, 20 July 2019, www.iflscience. com/125000yearold-bone-engravings-may-be-oldest-ever-example-of-denisovan-art-53134.

13 R. Wragg Sykes, *Kindred: Neanderthal life, love, death and art*, Bloomsbury Sigma, London, 2021; R. Wragg Sykes, 'The puzzle of Neanderthal aesthetics', 1 May 2023, www.bbc.com/future/article/20230428-the-puzzle-of-neanderthal-culture-and-aesthetics.

14 D. Leder, R. Hermann, M. Hüls, G. Russo, P. Hoelzmann, R. Nielbock, U. Böhner, J. Lehmann, M. Meier, A. Schwalb, A. Tröller-Reimer, T. Koddenberg & T. Terberger, 'A 51,000-year-old engraved bone reveals Neanderthals' capacity for symbolic behaviour', *Nature Ecology & Evolution*, vol. 5, 2021, pp. 1273–82.

15 Wragg Sykes, *Kindred*, pp. 244–46.

16 B. David, *Cave Art*, Thames & Hudson, London, 2017, pp. 9–10.

17 D. Lewis-Williams, *The Mind in the Cave: Consciousness and the origins of art*, Thames & Hudson, London, 2002; D. Lewis-Williams & D. Pearce, *Inside the Neolithic Mind: Consciousness, cosmos and the realm of the gods*, Thames & Hudson, London, 2009.

18 Bahn, *Images of the Ice Age*, pp. 287–92.

19 A. Paterson, 'San rain images in the Cederberg', *The Digging Stick*, vol. 35, no. 1, 2018, p. 22. The first paragraph references D. Lewis-Williams, *The Mind in the Cave*, p. 129.

20 J. Ross & I. Davidson, 'Rock art and ritual: An archaeological analysis of rock art in arid Central Australia', *Journal of Archaeological Method and Theory*, vol. 13, 2006, p. 308.

21 Davidson, 'Symbolism and becoming a hunter-gatherer', p. 7.

22 Bahn, *Images of the Ice Age*, p. 27.

23 A. Brumm, A.A. Oktaviana, B. Burhan, B. Hakim, R. Lebe, J.-X. Zhao, P.H. Sulistyarto, M. Ririmasse, S. Adhityatama, I. Sumantri & M. Aubert, 'Oldest cave art found in Sulawesi', *Science Advances*, vol. 7, no. 3, 2021, doi: 10.1126/sciadv.abd4648.

24 A. Rosenfeld, 'Rock-art as an indicator of changing social geographies in Central Australia', in B. David & M. Wilson (eds), *Inscribed Landscapes: Marking and making place*, University of Hawaii Press, Honolulu, 2002, pp. 61–78.

25 P.S.C. Taçon, S.K. May, U.K. Frederick & J. McDonald, *Histories of Australian Rock Art Research*, ANU Press, Canberra, 2022.

26 Morphy, *Ancestral Connections*, p. 265; Morphy, *Aboriginal Art*, pp. 97, 111–14.

27 M.J. Morwood, *Visions from the Past: The archaeology of Australian Aboriginal art*, Allen & Unwin, Crows Nest, 2002, p. 41.

28 Kelly, *Knowledge and Power in Prehistoric Societies*, pp. 72–73.

29 D.S. Whitley, *Introduction to Rock Art Research*, Left Coast Press, Walnut Creek, 2011, pp. 165–68.

30 Screen Australia, *First Footprints*, Ultimo, 2013. I cannot recommend this TV series more highly if you want to get an impression of the role of art and of place in Indigenous knowledge systems.

31 J. Delannoy, B. David, J. Geneste, M. Katherine, B. Barker, R.L. Whear & R.B. Gunn, 'The social construction of caves and rockshelters: Chauvet cave (France) and Nawarla Gabarnmang (Australia)', *Antiquity*, vol. 87, no. 335, 2013, pp. 12–29.

32 David, *Cave Art*, pp. 43–64.

33 David, *Cave Art*, p. 52.

34 B.W. Smith, J.L. Black, K.L. Mulvaney & S. Hœrlé, 'Monitoring rock art decay: Archival image analysis of petroglyphs on Murujuga, Western Australia', *Conservation and Management of Archaeological Sites*, vol. 23, no. 5–6, 2021, pp. 198–220.

35 K. Mulvaney, 'About time: Toward a sequencing of the Dampier Archipelago petroglyphs of the Pilbara region, Western Australia', *Records of the Western Australian Museum*, Supplement, vol. 79, 2011, pp. 30–49.

36 Examples redrawn from Mulvaney, 'About time', p. 42.

37 Portion of display labelled as 'Female figurines from Dolní Věstonice' from the Anthropos Pavilion, Moravské zemské muzeum, Brno, South Moravia, Czech Republic.

38 Bahn, *Images of the Ice Age*, pp. 60 and 216.

39 G. von Petzinger, *The First Signs*, Simon & Schuster, New York, 2016.

40 B. Bacon, A. Khatiri, J. Palmer, T. Freeth, P. Pettitt & R. Kentridge, 'An upper Palaeolithic proto-writing system and phenological calendar', *Cambridge Archaeological Journal*, vol. 33, no. 3, 2023, pp. 371–89.

41 Bahn, *Images of the Ice Age*, p. 197.

42 Bahn, *Images of the Ice Age*, p. 1.

43 Bahn, *Images of the Ice Age*, pp. 54–56.

44 S. Chisena & C. Delage, 'On the attribution of Palaeolithic artworks: The case of La Marche (Lussac-les-Châteaux, Vienne)', *Open Archaeology*, vol. 4, 2018, pp. 239–61.

45 Bahn, *Images of the Ice Age*, p. 62.

46 Kelly, *Knowledge and Power in Prehistoric Societies*, pp. 87–89.

47 Bahn, *Images of the Ice Age*, p. 139.

48 R. Lankester, *Secrets of Earth and Sea*, The Macmillan Company, New York, 1920, Figures 1.4 and 1.5.

49 M. Titiev, 'The story of Kokopele', *American Anthropologist*, vol. 41, no. 1, 1939, pp. 91–98.

50 Nooter Roberts & Roberts, *Luba*, pp. 12–13.

51 Lévi-Strauss, *The Savage Mind*, p. 154.

52 Bahn, *Images of the Ice Age*, p. 204.

53 W. Herzog, *Cave of Forgotten Dreams*, 2010.

54 Delannoy et al., 'The social construction of caves and rockshelters', p. 15.

55 M. Azéma & F. Rivère, 'Animation in Palaeolithic art: A pre-echo of cinema', *Antiquity*, vol. 86, no. 332, 2012, pp. 316–24; M. Azéma, 'Animation and graphic narration in the Aurignacian', *Palethnologie*, vol. 7, 2015, http://journals.openedition.org/palethnologie/861.

56 S.J. Waller, 'Hear here: Prehistoric artists preferentially selected reverberant spaces and choice of subject matter underscores ritualistic use of sound', in L. Büster, E. Warmenbol & D. Mlekuž (eds), *Between Worlds: Understanding ritual cave use in later prehistory*, Springer, New York, 2018, p. 260.

57 An excellent coverage of the discovery of the artworks globally can be found in David, *Cave Art*, pp. 16–45.

58 Bahn, *Images of the Ice Age*, pp. 62, 180, 195.

59 K. Helskog, 'Landscapes in rock-art: Rock-carving and ritual in the old European North', in C. Chippindale & G. Nash (eds), *Pictures in Place: The figured landscapes of rock-art*, Cambridge University Press, Cambridge, 2004, pp. 265–88.

60 J. Iriarte, M.J. Ziegler, A.K. Outram, M. Robinson, P. Roberts, F.J. Aceituno, G. Morcote-Ríos & T.M. Keesey, 'Ice Age megafauna rock art in the Colombian Amazon?', *Philosophical Transactions of the Royal Society*, B377, 2022, 20200496, doi: 10.1098/rstb.2020.0496.

61 J.D. Spangler & I. Davidson, 'Finding order out of chaos: A statistical analysis of Nine Mile Canyon rock art', in I. Davidson & A. Nowell (eds), *Making Scenes: Global perspectives on scenes in rock art*, Berghahn Books Inc., New York, 2021, pp. 277–94.

62 L.V. Benson, E.M. Hattori, J. Southon & B. Aleck, 'Dating North America's oldest petroglyphs, Winnemucca Lake subbasin, Nevada', *Journal of Archaeological Science*, vol. 40, no. 12, pp. 4466–76.

63 K.F. Anschuetz, 'A healing place: Rio Grande Pueblo cultural land-scapes and the Petroglyph National Monument', in K.F. Anschuetz (ed.), *'That place people talk about': The Petroglyph National Monument ethnographic*

landscape report, unpublished manuscript, on file with National Park Service, Petroglyph National Monument, Albuquerque, 2002, pp. 3.1–3.47.

64   I.N.M. Wainwright, H. Sears & S. Michalski, 'Design of a rock art protective structure at Petroglyphs Provincial Park, Ontario, Canada', *Journal of the Canadian Association for Conservation*, vol. 22, 1997, pp. 53–76.

65   See 'Tassili n'Ajjer', Unesco World Heritage Convention, https://whc.unesco.org/en/list/179.

66   Panel adapted from A.F.C. Holl, *Saharan Rock Art: Archaeology of Tassilian Pastoralist Iconography*, African Archaeology Series, AltaMira Press, Walnut Creek, 2004, pp. 115, 127.

67   The Bradshaw Foundation, www.bradshawfoundation.com.

68   F.J. Korom, *Village of Painters: Narrative scrolls from West Bengal*, Museum of New Mexico Press, Santa Fe, 2006.

69   R.O. Roberts, R.H. Cha, M.M. Mielke, Y.E. Geda, B.F. Boeve, M.M. Machulda, D.S. Knopman & R.C. Petersen, 'Risk and protective factors for cognitive impairment in persons aged 85 years and older', *Neurology*, vol. 84, no. 18, 2015, pp. 1854–61.

70   'Creating art could help delay the onset of dementia', *Paintings in Hospitals*, n.d., www.paintingsinhospitals.org.uk/46-creating-art-could-help-delay-the-onset-of-dementia.

Chapter 9: The monumental story of knowledge spaces

1   Kelly, *Knowledge and Power in Prehistoric Societies*, pp. 1–12, 165–237; Kelly, *The Memory Code*, pp. 80–294.

2   T. Terberger, M. Zhilin & S. Savchenko, 'The Shigir Idol in the context of early art in Eurasia', *Quaternary International*, vol. 573, 2021, pp. 14–29.

3   UNESCO World Heritage Site: Göbekli Tepe, see https://whc.unesco.org/en/list/1572; E.B. Banning, 'So fair a house: Gobekli Tepe and the identification of temples in the pre-pottery Neolithic of the Near East', *Current Anthropology*, vol. 52, no. 5, 2011, pp. 619–60; A. Curry, 'Gobekli Tepe: The world's first temple?', *The Smithsonian*, November 2008, www.smithsonianmag.com/history-archaeology/gobekli-tepe.html; S. Scham, 'The world's first temple', *Archaeology*, vol. 61, no. 6, 2008, pp. 22–27.

4   A. Lorenzis & V. Orofino, 'New possible astronomic alignments at the megalithic site of Göbekli Tepe, Turkey', *Archaeological Discovery*, vol. 3, no. 1, 2015, pp. 40–50.

5   Spitzer, *The Musical Human*, p. 158.

6   E. Özdoğan, 'The Sayburç reliefs: A narrative scene from the Neolithic', *Antiquity*, vol. 96, no. 390, 2022, p. 1599.

7   Spitzer, *The Musical Human*, pp. 159–60; Morley, *The Prehistory of Music*, p. 121.

8 A.M. David, M. Cole, T. Horsley, N. Linford, P. Linford & L. Martin, 'A rival to Stonehenge? Geophysical survey at Stanton Drew, England', *Antiquity*, vol. 78, no. 300, 2004, pp. 341–58; J. Oswin, *Stanton Drew 2010: Geophysical survey and other archaeological investigations*, Bath and Camerton Archaeological Society, Bath, 2011, p. 29; N. Linford, *Stanton Drew Stone Circles and Avenues, Stanton Drew, Bath and North-East Somerset*, Research Report Series no. 79, 2017, Historic England, Portsmouth.

9 My images of Stanton Drew have been adapted from a number of sites, including Google Maps, English Heritage (www.english-heritage.org.uk/visit/places/stanton-drew-circles-and-cove) and Rupert Soskin's images in his 2020 film, *Standing with Stones* (www.youtube.com/watch?v=Iq4xM8TLWc0).

10 Oswin, *Stanton Drew 2010*, p. 29.

11 Oswin, *Stanton Drew 2010*, p. 38.

12 Soskin, *Standing with Stones*, 2020, www.youtube.com/watch?v=Iq4xM8TLWc0.

13 Kelly, *Knowledge and Power in Prehistoric Societies*, pp. 203, 208, 223, 230; Kelly, *The Memory Code*, pp. 152–65.

14 C. Wickham-Jones, *Between the Wind and the Water: World Heritage Orkney*, Windgather Press, Macclesfield, 2006.

15 A. Watson & D. Keating, 'The architecture of sound in Neolithic Orkney', in A. Ritchie (ed.), *Neolithic Orkney in Its European Context*, McDonald Institute for Archaeological Research, Cambridge, 2000, pp. 259–63.

16 N. Card, J. Cluett, J. Downes, J. Gater & S. Ovenden, 'The heart of Neolithic Orkney World Heritage Site: Building a landscape', in M. Larsson & M. Parker Pearson (eds), *From Stonehenge to the Baltic: Living with cultural diversity in the third millennium BC*, Archaeopress, London, 2007, pp. 221–29.

17 D.N. Marshall, 'Carved stone balls', *Proceedings of the Society of Antiquaries of Scotland*, vol. 108, 1977, pp. 40–72; G. MacGregor, 'Making sense of the past in the present: A sensory analysis of carved stone balls', *World Archaeology*, vol. 31, no. 2, 1999, pp. 258–71.

18 C. Ruggles & G. Barclay, 'Cosmology, calendars and society in Neolithic Orkney: A rejoinder to Euan MacKie', *Antiquity*, vol. 74, no. 283, 2000, p. 62.

19 V. Cummings & A. Pannett, 'Island views: The settings of the chambered cairns of southern Orkney', in V. Cummings & A. Pannett (eds), *Set in Stone: New approaches to Neolithic monuments in Scotland*, Oxbow, Oxford, 2005, pp. 14–24; E.W. MacKie, 'Maeshowe and the winter solstice: Ceremonial aspects of the Orkney grooved ware culture', *Antiquity*, vol. 71, no. 272, 1997, pp. 338–59.

20 Kelly, *The Memory Code*, pp. 166–82.

21 G. Stout & M. Stout, *Newgrange*, Cork University Press, Cork, 2008.

22 J. Thomas, *Understanding the Neolithic* (2nd edn), Routledge, London, 1999, pp. 47, 139, 207; V. Cummings, 'The architecture of monuments', in J. Pollard (ed.), *Prehistoric Britain*, Blackwell, Oxford, 2008, p. 138.

23 My detailed academic and mainstream analysis of the details of Stonehenge, Avebury and the broader British context is in Kelly, *Knowledge and Power in Prehistoric Societies*, pp. 203–33. An extensive bibliography is available from my website, www.lynnekelly.com.au. For a mainstream audience, I detailed the archaeology of British Neolithic sites and my interpretation in Kelly, *The Memory Code*, pp. 98–151.

24 A. Burl, *Prehistoric Avebury* (2nd edn), Yale University Press, New Haven, 2002, p. 70.

25 D.J. Nash, T.J.R. Ciborowski, J.S. Ullyott, M.P. Pearson, T. Darvill, S. Greaney, G. Maniatis & K.A. Whitaker, 'Origins of the sarsen megaliths at Stonehenge', *Science Advances*, vol. 6, no. 31, 2020, doi: 10.1126/sciadv.abc0133.

26 R. Till, 'Songs of the stones: An investigation into the acoustic culture of Stonehenge', *Journal of the International Association for the Study of Popular Music*, vol. 1, no. 2, 2010, pp. 1–18; T.J. Cox, M. Bruno, B.M. Fazenda & S.E. Greaney, 'Using scale modelling to assess the prehistoric acoustics of Stonehenge', *Journal of Archaeological Science*, vol. 122, 2020, 105218, doi: 10.1016/j.jas.2020.105218; Spitzer, *The Musical Human*, p. 161.

27 T. Darvill, 'Keeping time at Stonehenge', *Antiquity*, vol. 96, no. 386, 2022, pp. 319–35.

28 See Kelly, *Knowledge and Power in Prehistoric Societies*, pp. 215, 217, 224–6; Kelly, *The Memory Code*, pp. 130, 132–4.

29 See Kelly, *Knowledge and Power in Prehistoric Societies*, pp. 10, 29–30, 227–8; Kelly, *The Memory Code*, pp. 26, 217, 224–6.

30 A. Burl, *A Brief History of Stonehenge: A complete history and archaeology of the world's most enigmatic stone circle*, Robinson, London, 2007, p. 27.

31 Caesar, as quoted in P.B. Ellis, *Caesar's Invasion of Britain*, New York University Press, New York, 1980, p. 30; M.J. Green, 'The time lords: Ritual calendars, druids and the sacred year', in D.D.A. Simpson & A.M. Gibson (eds), *Prehistoric Ritual and Religion*, Sutton Publishing, Stroud, 1998, p. 190.

32 Kelly, *The Memory Code*, pp. 183–97.

33 C. Scarre, *Landscapes of Neolithic Brittany*, Oxford University Press, Oxford, 2011, p. 100; M. Patton, *Statements in Stone: Monuments and society in Neolithic Brittany*, Routledge, London, 1993, p. 62.

34 P. Bueno Ramírez, R. Barroso Bermejo & R. de Balbin Behrmann, 'Pigments for the dead: Megalithic scenarios in southern Europe', *Archaeological and Anthropological Sciences*, vol. 15, 2023, doi: 10.1007/s12520-023-01850-0.

35 Kelly, *The Memory Code*, pp. 220–36.

36 A detailed and scientifically rigorous book on the Nazca Lines is A.F. Aveni, *Between the Lines: The mystery of the giant ground drawings of ancient Nasca, Peru*, University of Texas Press, Austin, 2000.

37 A. Aveni & H. Silverman, 'Between the lines: Reading the Nazca markings as rituals writ large', in A.F. Aveni (ed.), *Foundations of New World Cultural Astronomy: A reader with commentary*, University Press of Colorado, Boulder, 2008, pp. 621–33.

38 The Brooklyn Museum website has an excellent record for the Paracas Textile, allowing you to see the figures in detail. The description is extensive and fascinating. See www.brooklynmuseum.org/opencollection/objects/48296.

39 M. Campbell, 'Memory and monumentality in the Rarotongan landscape', *Antiquity*, vol. 80, no. 307, 2006, p. 115.

40 Papahānaumokuākea Marine National Monument, www.papahanaumo kuakea.gov/monument_features/cultural_mmm_upright.html.

41 'Mokumanamana Ki'i Pōhaku', Bishop Museum, https://hawaiialive.org/mokumanamana-kii-pohaku.

42 'Ki'i', National Park Service, 2 May 2020, www.nps.gov/puho/learn/historyculture/kii.htm.

43 B. Haami, *Pūtea Whakairo: Māori and the written word*, University of Hawaii Press, Honolulu, 2006.

44 I have written before, in a great deal more detail, about Rapa Nui; see Kelly, *The Memory Code*, pp. 271–94. The main references used were: J. Flenley & P. Bahn, *The Enigmas of Easter Island: Island on the edge* (2nd edn), Oxford University Press, Oxford, 2003; T. Hunt & C. Lipo, *The Statues that Walked: Unraveling the mystery of Easter Island*, Free Press, New York, 2011; and K. Routledge, *The Mystery of Easter Island*, Rapa Nui Press, Rapa Nui, 2005 [1919].

45 Routledge, *The Mystery of Easter Island*, p. 171.

46 The 'magnificent avenues' are wonderfully detailed in Routledge, *The Mystery of Easter Island*, p. 196.

47 Kelly, *Knowledge and Power in Prehistoric Societies*, pp. 186–202; Kelly, *The Memory Code*, pp. 240–49.

48 J. Ellerbe & D.M. Greenlee, *Poverty Point: Revealing the forgotten city*, Louisiana State University Press, Baton Rouge, 2015.

49 C. Carbone, 'Mount Horeb Earthwork', Society of Architectural Historians, https://sah-archipedia.org/buildings/KY-01-067-0057.

50 Excavation reports by archaeologist Robert Riordan and the masters thesis of Katherine Lynn Rippl, 'Examination of two post circles found in the Ohio Valley', Michigan State University (personal contact).

51 T.R. Pauketat & S.M. Alt, 'Mounds, memory, and contested Mississippian history', in R.M. Van Dyke & S.E. Alcock (eds), *Archaeologies of Memory*, Blackwell, Malden, 2003; T.R. Pauketat, *Ancient Cahokia and the Mississippians*, Cambridge University Press, Cambridge, 2004.

52 W. Iseminger, *Cahokia Mounds: America's first city*, The History Press, Charleston, 2010.

53 J.Z. Holt, 'Rethinking the Ramey state: Was Cahokia the center of a theater state?', *American Antiquity*, vol. 74, no. 2, 2009, pp. 231–54; J.Z. Holt, 'Ritual objects and the Red Horn state: Decoding the theater state at Cahokia', *Illinois Antiquity*, vol. 48, no. 3, 2013, pp. 17–19.

54 Kelly, *Knowledge and Power in Prehistoric Societies*, pp. 146–85; Kelly, *The Memory Code*, pp. 200–19. Among many references on Chaco Canyon, this section is informed in particular by S.H. Lekson, *The Chaco Meridian: Centers of political power in the ancient Southwest*, AltaMira Press, Walnut Creek, 1999; D.G. Noble (ed.), *In Search of Chaco: New approaches to an archaeological enigma*, School of American Research Press, Santa Fe, 2004; R.M. Van Dyke, *The Chaco Experience: Landscape and ideology at the Center Place*, School for Advance Research Press, Santa Fe, 2007; R.M. Van Dyke, 'Memory and the construction of Chacoan society', in R.M. Van Dyke & S.E. Alcock, *Archaeologies of Memory*, Blackwell, Malden, 2003.

55 L.J. Kuwanwisiwma, 'Yupköyvi: The Hopi story of Chaco Canyon', in D.G. Noble (ed.), *In Search of Chaco*, pp. 41–8.

Chapter 10: The impact of literacy on our knowledge gene skills

1 This chapter draws on many references, but hugely influential was A. Robinson, *The Story of Writing*, Thames & Hudson, London, 1995. The dates are from page 12 and the ideas on derived scripts from page 45.

2 I have written in more detail about Mesoamerican cultures in Kelly, *The Memory Code*, pp. 237–70. A full bibliography is available on my website, www.lynnekelly.com.au.

3 F. Meddens & M. Frouin, 'Inca sacred space, platforms and their potential soundscapes, preliminary observations at usnu from Ayacucho', *Revista Haucaypata*, vol. 1, 2011, pp. 24–40.

4 G. Brokaw, *A History of the Khipu*, Cambridge University Press, Cambridge, 2010, pp. 2–21.

NOTES

5   Robinson, *The Story of Writing*, pp. 121–30.

6   The Scala Group (ed.), *Precolumbian Art: The pocket visual encyclopedia of art*, Welcome Rain Publishers, New York, 2013.

7   Spitzer, *The Musical Human*, p. 166–69.

8   D. Tedlock (translator), *Popol Vuh: The definitive edition of the Mayan book of the dawn of life and the glories of gods and kings*, Simon & Schuster, New York, 1996.

9   G. Shelach-Lavi, *The Archaeology of Early China: From prehistory to the Han Dynasty*, Cambridge University Press, Cambridge, 2015.

10 X. Li, G. Harbottle, J. Zhang & C. Wang, 'The earliest writing? Sign use in the seventh millennium BC at Jiahu, Henan Province, China', *Antiquity*, vol. 77, no. 295, 2003, pp. 31–44.

11 K. Wilson & J. Douglas, 'In Sync: Ancient Chinese bronze bells at the Smithsonian', *Smithsonian Music*, March 2016, https://music.si.edu/story/sync-ancient-chinese-bronze-bells-smithsonian.

12 Robinson, *The Story of Writing*, pp. 183–84.

13 S. Sheng, 'Huashan Mountain petroglyphs', in C. Smith (ed.), *Encyclopedia of Global Archaeology*, Springer, New York, 2014.

14 I.F. Gelati-Meinert, P. Robinson & T. Zhu, *Chinese Folk Performing Art— Chinese Culture Series*, China Pictorial Publishing House, Beijing, 2014.

15 D. Moser, *A Billion Voices: China's search for a common language*, Penguin Books, Melbourne, 2016.

16 T.O. Höllmann, *Chinese Script: History, characters, calligraphy*, Columbia University Press, New York, 2017, p. 91.

17 Robinson, *The Story of Writing*, pp. 199–203.

18 D.C.S. Li, R. Aoyama & T. Wong, 'Silent conversation through Brushtalk (筆談): The use of Sinitic as a scripta franca in early modern East Asia', *Global Chinese*, vol. 6, no. 1, 2020, pp. 1–24.

19 B. Koyama-Richard, *One Thousand Years of Manga* (rev. edn), Thames & Hudson, New York, 2022, p. 64.

20 Robinson, *The Story of Writing*, pp. 71, 91.

21 Robinson, *The Story of Writing*, pp. 80, 83.

22 A sample cuneiform tablet recording music can be seen in the collection of the Penn Museum at www.penn.museum/collections/object/527649, Tablet Fragment N3354, dated to 1900–1600 BCE. The analysis of the musical inscriptions can be found in A.D. Kilmer & M. Civil, 'Old Babylonian musical instructions relating to hymnody', *Journal of Cuneiform Studies*, vol. 38, no. 1, 1986, pp. 94–98.

23 Robinson, *The Story of Writing*, pp. 46, 93–94.

24 Robinson, *The Story of Writing*, p. 165.

25 E.A. Havelock, *Preface to Plato*, Harvard University Press, Cambridge, 1963, pp. 93–94.

26 F.A. Yates, *The Art of Memory*, Routledge and Kegan Paul, London, 1966.

27 M.L. Mark, 'The evolution of music education philosophy from utilitarian to aesthetic', *Journal of Research in Music Education*, vol. 30, no. 1, 1982, pp. 15–21.

28 H.A. Innis, *The Bias of Communication*, University of Toronto Press, Toronto, 1964, pp. 109–10.

29 A.B. Lord, *The Singer of Tales*, Harvard University Press, Cambridge, 1960.

30 T. Gioia, *Music: A subversive history*, Basic Books, New York, 2019, p. 87.

31 I have written about medieval memory systems in more detail in Kelly, *Memory Craft*, pp. 11–30.

32 I have created a bestiary for remembering names and words in general. I have another for Chinese (Mandarin) vocabulary, and a third for French. The bestiaries, and instructions on how to use them, are freely available from my website: www.lynnekelly.com.au.

33 Jacobus Publicius, as quoted in M. Carruthers & J.M. Ziolkowski (eds), *The Medieval Craft of Memory: An anthology of texts and pictures*, University of Pennsylvania Press, Philadelphia, 2004, p. 237.

34 Robinson, *The Story of Writing*, p. 34.

Chapter 11: Putting our knowledge gene to work

1 I have written in detail about setting up memory palaces in Kelly, *Memory Craft*, pp. 31–60.

2 D.J. Levitin, J.A. Grahn & J. London, 'The psychology of music: Rhythm and movement', *Annual Review of Psychology*, vol. 69, 2018, pp. 51–75.

3 Harvey, *Music, Evolution, and the Harmony of Souls*, especially p. 177.

4 A. Collins, *The Music Advantage: How music helps your child's brain*, Allen & Unwin, Crows Nest, 2020, p. 35.

5 Collins, *The Music Advantage*, p. 62.

6 J. Shand, 'Dark leads us back from the live music wilderness with Brel', *The Sydney Morning Herald*, 7 November 2021.

7 Gioia, *Music*, p. 87. The rest of the ideas come from our personal correspondence, but can also be found in the book.

8 Mike Hazard, *Wiigwaasabakoon/Birch Bark Scrolls*, 2011, www.youtube.com/watch?v=lfSB4uyhMUk.

9 I have written in detail about using story to remember multiplication tables in Kelly, *Memory Craft*, pp. 210–14.

INDEX

Page and plate numbers in *italics* refer to illustrations

INDEX